Anthocyanin

Purple fruit e

Black Currents
Blue berrie

red cabbage
Aubergene
Puple sweet potatue

CW00524458

PHYTOCHEMICALS

HOW TO ORDER THIS BOOK

BY PHONE: 800-233-9936 or 717-291-5609, 8AM–5PM Eastern Time

BY FAX: 717-295-4538

BY MAIL: Order Department
Technomic Publishing Company, Inc.
851 New Holland Avenue, Box 3535
Lancaster, PA 17604, U.S.A.

BY CREDIT CARD: American Express, VISA, MasterCard

BY WWW SITE: http://www.techpub.com

PERMISSION TO PHOTOCOPY–POLICY STATEMENT

Authorization to photocopy items for internal or personal use, or the internal or personal use of spe-
cific clients, is granted by Technomic Publishing Co., Inc. provided that the base fee of US $3.00 per
copy, plus US $ .25 per page is paid directly to Copyright Clearance Center, 222 Rosewood Drive,
Danvers, MA 01923, USA. For those organizations that have been granted a photocopy license by
CCC, a separate system of payment has been arranged. The fee code for users of the Transactional
Reporting Service is 1-56676/98 $5.00 + $ .25.

PHYTOCHEMICALS
— *A New Paradigm* —

Edited by

Wayne R. Bidlack, Ph.D.
College of Agriculture
California State Polytechnic University, Pomona

Stanley T. Omaye, Ph.D., F.A.T.S.
Department of Nutrition
University of Nevada, Reno

Mark S. Meskin, Ph.D., R.D.
Department of Food, Nutrition and Consumer Science
California State Polytechnic University, Pomona

Debra Jahner, M.S.
Senior Research Scientist
Rehnborg Center for Nutrition and Wellness
Nutrilite Division of Amway Corporation

LANCASTER · BASEL

Phytochemicals

aTECHNOMIC publication

Technomic Publishing Company, Inc.
851 New Holland Avenue, Box 3535
Lancaster, Pennsylvania 17604 U.S.A.

Copyright ©1998 by Technomic Publishing Company, Inc.
All rights reserved

No part of this publication may be reproduced, stored in a
retrieval system, or transmitted, in any form or by any means,
electronic, mechanical, photocopying, recording, or otherwise,
without the prior written permission of the publisher.

Printed in the United States of America
10   9   8   7   6   5   4   3   2   1

Main entry under title:
  Phytochemicals: A New Paradigm

A Technomic Publishing Company book
Bibliography: p.
Includes index p. 177

Library of Congress Catalog Card No. 98-86874
ISBN No. 1-56676-684-2

Contents

Preface

As a complete mixture of chemicals, food provides essential nutrients, requisite calories, and other physiologically active constituents needed for life and health. Except for a few nutrients, most of these dietary chemicals remain uncharacterized, including the level and frequency of human consumption. Many of these chemicals are inert, some are toxic or carcinogenic, while others may have positive effects on physiologic function acting as protective agents countering the risks of acute toxicity and diminishing the onset of disease. Further evaluation of the mechanisms of action of these chemicals is essential to our understanding of why certain chemicals seem to be beneficial at low concentrations within our general food supply.

Epidemiologic surveys have provided positive correlations between certain diets and specific foods. Diets rich in fruits, vegetables, and grains have been associated with the prevention of several chronic diseases. There has been no evolutionary pressure exerted on plants to develop food components that protect man from diseases such as cancer, yet diets rich in fruits and vegetables appear to do just that. Most likely, the active plant constituents developed as a part of their own defense mechanisms and only fortuitously contribute to man's health.

A new diet-health paradigm may be evolving that acknowledges the nutritious and healthful aspects of food but goes beyond the role of food constituents as essential nutrients for sustaining life and growth to one preventing or delaying premature onset of chronic diseases later in life. The key change in the new health paradigm is prevention rather than treatment.

The promise of functional foods (Bidlack and Wang, 1998) has emerged at a time when consumer interest in diet and health is at an all-time high. Functional foods, food products, and supplements that deliver a possible physiologic benefit in the management or prevention of disease represent an opportunity for future new product growth in the food and beverage industries.

Researchers are examining these foods, isolating and characterizing chemical components, structures, and physiologic function. The number of physiologically active food components has increased dramatically in the last decade. Possible naturally occurring protective agents include the carotenoids, vitamins C and E, the flavonoids and other phenolic compounds, phytoestrogens, indoles, and fiber. The extent of experiments needed to thoroughly characterize the physiologically active phytochemicals is described in Table 1 (Bidlack and Wang, 1998). To date, very few phytochemicals have been examined as thoroughly as needed. It will be many years before these agents are determined to offer health benefits as either natural ingredients, food additives, or as dietary supplements. Decisions must be based on efficacy, assessing the lowest possible dose to provide the benefit. In all cases, safety must be assured.

Epidemiologic findings, supported by animal studies, have led to recommendations that American should consume at least two servings of fruit and three servings of vegetables daily. The majority of adults falls well short of meeting these guidelines.

The relation between diet and human cancer risk is a relatively recent phenomenon. Diet ranks second only to smoking as a leading contributor to cancer incidence and mortality. In chapter 1, Swanson describes the epidemiologic evidence that diets rich in fruits and vegetables are associated with reduced risk of a variety of tumors, especially epithelial cancers of the respiratory and gastrointestinal tract. She indicates that the effect of whole fruit and vegetables tends to be more pronounced than that of the individual dietary constituents they supply. Perhaps, it is the combination of nutrients and other dietary constituents in fruit and vegetables that is etiologically relevant.

While essential nutrients, particularly the antioxidants, have been studied extensively, numerous "nonessential" dietary plant constituents (e.g., phenolics, organosulfur compounds, flavonoids, and isoflavones) clearly have anticarcinogenic potential. Unlike the essential nutrients, food composition data for most nonessential phytochemicals either do not exist or are limited. Thus, epidemiologic data for most bioactive plant constituents are sparse.

In general, the effect of whole foods is more pronounced than that of the individual micronutrients or other bioactive phytochemicals they supply. Sorting out the possible protective components in fruits and vegetables has

TABLE 1. **Several Areas of Research Need to Be Investigated to Better Define the Role of Phytochemicals in Health.**

1. Identify the specific types of phytochemicals that provide health benefits
   - determine the strength of association
   - characterize the sources, diets or supplements, of phytochemicals that are beneficial or harmful
2. Characterize the factors that affect absorption and bioavailability of the phytochemicals
3. Determine the metabolic fate of absorbed phytochemicals
4. Establish the levels of phytochemicals identified with specific tissues
   - determine specific functions of the phytochemicals in these tissues
   - identify the existence of specific binding proteins
   - identify selective uptake mechanisms
   - determine species specificity
   - identify differences in metabolic pathways in tissues accumulating different forms
5. Identify and characterize metabolites of phytochemical metabolism
   - determine physiologic activities of metabolic products
6. Establish safety of use
   - determine the concentrations at which pharmacological doses become a toxicological problem
7. Determine the type of phytochemical and the effective dose that protect against disease, e.g., cancer, coronary heart disease, diabetes, and osteoporosis
8. Define the saturation point of phytochemicals that provide protection against cancer
   - determine dose response
   - determine effect of intervention on precancerous stage vs existing tumors
   - evaluate chemically induced models vs spontaneous tumor models
   - evaluate timing of dose to the onset of cancer
9. Identify new mechanisms by which the phytochemicals produce protective effects
   - gap junction communication
   - immunomodulation
   - cell differentiation
10. Characterize the effects of phytochemicals at various concentrations and using specific isomers
11. Determine optimal phytochemical mixtures
    - determine composition
    - duration of feeding
    - amounts to be fed
12. Identify the proportion of the population likely to respond positively to phytochemicals
13. Establish the pharmacokinetics of delivered dose
    - evaluate single and combined doses
    - evaluate with and without food sources present
14. More closely examine the dietary components associated with health and disease prevention from the diet as a whole

Modified from Bidlack and Wang (1998).

been extremely difficult, but numerous "nonessential" dietary plant constituents clearly have anticarcinogenic potential.

While there is uncertainty as to which dietary phytochemicals or combinations of plant constituents confer protection, the significance of the role oxidative stress plays in the development of certain chronic diseases has been recognized. The ability of antioxidants to retard the processes of oxidation has led to the hypothesis that ingestion of such nutrients may prevent these disease states. In Chapter 2, German and Dillard discuss the balance between the known benefits and risks of free radical oxidation chemistry, noting they can be shifted in either a detrimental or beneficial direction by changing the dominance of the pro-oxidative processes and antioxygenic reactions. Nonnutrient components of the diet actively participate in the oxidative balance of living cells. The process of oxidative chemistry is multistage, and its complexity forms the basis for multiple mechanisms of pathogenesis. Further research needs to be implemented on specific molecules, mechanisms of action, biomarkers of their status and functions, efficacious delivery, and means to improve the food supply to realize the full value of the antioxidant.

α-tocopherol has also served to inhibit free radical reactions in biological membranes. With the discovery of its vitamin activity, most research has focused on this single compound of the tocopherol family. However, researchers have always thought that vitamin E may have additional roles in metabolism. In Chapter 3, Hood suggests that these other functions may reside with tocotrienols.

The tocotrienols have an excellent antioxidant profile. In low doses, dietary tocotrienols inhibit HMGCoA reductase, resulting in a lowering of serum cholesterol. Tocopherols do not demonstrate this function. In addition, tocotrienols have been shown to delay the onset of tumorigenesis and/or reduce the proliferation of a range of tumors.

Little knowledge exists about the mechanism of action of phytochemicals or their role in health promotion benefit and disease prevention. Even with the more studied chemicals, such as β-carotene, tocopherol, and ascorbic acid, neither the basic function at the molecular level nor the intricate interactions that occur between these components and others in the diet are understood.

In most cases, basic studies evaluate individual component reactions while ignoring complex interactions between the components. In Chapter 4, Omaye and Zhang present a new experimental view that requires an evaluation of interactions between phytochemicals, using β-carotene, α-tocopherol, and ascorbic acid as the model.

Antioxidant protection involves endogenous compounds, including various enzymes, and exogenous factors, of which many might be derived from dietary phytochemicals. It is apparent that, because of extensive interaction, both synergistic and antagonistic, that occurs *in vivo*, it can no longer be assumed that the outcome of their total action is equal to the sum of the antioxidant entity alone. The intake of excessive amounts of β-carotene and perhaps other carotenoids may lead to reduced vitamin E levels rather than a synergistic response. Studies on a single antioxidant may be misleading if expected interactions are not considered. Thus, multiple combinations of antioxidants and phytochemicals need to be studied.

Although the number of dietary carotenoids in fruits and vegetables is in excess of 40, to date only 13 have been identified as being absorbed, metabolized, and utilized by humans. Carotenoids, such as α- and β-carotene, are absorbed intact and converted to vitamin A. In contrast, metabolic transformations of other major carotenoids, such as lutein, zeaxanthin, and lycopene, involve a series of oxidation-reduction reactions.

In addition to the 13 carotenoids absorbed by humans, 13 of their stereoisomers and 8 of their metabolites have also been characterized in human serum and breast milk, employing a combination of two high-performance liquid chromatographic (HPLC) methods. This brings the total number of serum and milk carotenoids to 34 (characterized by Khachik in chapter 5).

Dietary carotenoids found in human organs and tissues such as liver, lung, breast, and cervix have also been identified and quantitated. Similarly, in addition to the presence of lutein and zeaxanthin in human retina reported previously, several oxidation products of these compounds were also identified. Geometrical isomers of lutein and zeaxanthin were also detected in human retina at low concentrations. Khachik proposed metabolic pathways of carotenoids, and their potential role in prevention of cancer and age-related macular degeneration is discussed.

Carotenoids may be differentiated according to the presence or absence of oxygen as part of their molecular structure. Oxycarotenoids contain one or more oxygen functions, while the remainder are hydrocarbons termed carotenes, including β-carotene. In Chapter 6, White and Paetau described the results of recent research evaluating the absorption and distribution of β-carotene and canthaxanthin, a model oxycarotenoid, in normolipidemic premenopausal women and examined subfractions of plasma triacylglycerol-rich lipoproteins. β-Carotene reduced the appearance of canthaxanthin in plasma, chylomicrons, and each very-low density lipoprotein subfraction. Canthaxanthin accumulated rapidly in low density lipoproteins (LDLs), and was not significantly affected by β-carotene. Canthaxanthin did not have a consistent effect on the appearance of β-carotene in plasma, plasma triacylglycerol-rich lipoproteins, or LDL. The results suggest distinct mechanisms of incorporation into lipoproteins and specific interactions of β-carotene and oxycarotenoids during intestinal absorption in humans.

The biflagellate unicellular alga *Dunaliella* serves as a natural source of β-carotene. In Chapter 7, Ben-Amotz describes the cultivation of autotrophic

Dunaliella in large-scale open ponds composed of salt, inorganic nutrients, carbon dioxide, and plenty of solar energy. The process yields β-carotene-rich algae ready for concentration by harvesting for use as high β-carotene algal powder or for extract β-carotene in oil.

The algal β-carotene differs from the synthetically available β-carotene in its stereoisomeric composition having equal amounts of all *trans* and 9 *cis* isomers and differing in their physicochemical features and antioxidative activity. The biotechnological production of natural β-carotene by *Dunaliella* and its unique stereoisomeric controlled composition has the potential to provide high-quality β-carotene for a wide variety of marketing applications.

Numerous organosulfur and organoselenium components of garlic and onions have been examined for health benefits. In Chapter 8, Block describes a variety of compounds in garlic and onion characterized by HPLC and coupled liquid chromatography-mass spectrometry (MS). Trace levels of organoselenium compounds also occur in *Allium* spp. These compounds, or derived products, are best identified using gas chromatography-atomic emission detection for volatiles and HPLC-inductively coupled plasma-MS for nonvolatiles. The initial chemistry induced by cutting garlic and onion, and subsequent changes caused by cooking or processing in the manufacture of commercial garlic nutritional supplements, must be considered when examining the molecular basis for the antibiotic and cancer-preventative activity as well as the beneficial effects on the cardiovascular system attributed to these plants or to commercial supplements produced from the plants.

As a unique area of interest, in Chapter 9, Stoner introduces potential applications for fungal components as food ingredients and supplement products. Historical, evolutionary, and ecological foundations for biochemical interactions among phytochemistry, fungi, and people point to important applications of fungi in food and pharmaceuticals. Unique biochemical capabilities, healthful and savory qualities, technical utility, and other advantages of specific fungi are presented. Medicinal chemistry of *Lentinus edobes* (Shiitake), of *Ganoderma lucidum* (LingZhi or Reishi), and of *Cordyceps* sp. (Chan Hua, Semitake, Tochukaso) is presented.

In Chapter 10, Litov provides a brief overview of the procedures needed to develop claims for new phytochemical products. Three regulatory categories were identified under which new phytochemical products may be regulated: (1) food, (2) dietary supplement, and (3) medical food. Food "health claims" are the most restrictive health messages, requiring the greatest commitment of time and resources for a new claim approval. Dietary supplement "nutritional support statements' have the most flexible language but are the least persuasive. Medical food "dietary management claims" are the most compelling and need significant clinical research support, but the products have limited application. To maximize the effectiveness of health messages, adequate research needs to be conducted to convince consumers of the link between a phytochemical and its health benefit.

"Natural products" are not a cure-all, but they have provided opportunity to identify phytochemicals that may have health-supporting properties without being nutrients. As a new paridigm in nutrition, perhaps some of these nonessential compounds will prove to be beneficial to health. The early example of this potential was the development of dietary fiber as a protective agent in gastrointestinal function.

An understanding of the basic principles of formulation, food technology and processing, and new biotechnologies provides ample opportunity for the development of functional foods, which will utilize phytochemicals having bioactive components to create products to prevent disease and maintain a healthy life throughout our existence. The future of science and product development in this area will be an exciting adventure for years to come.

WAYNE R. BIDLACK
Editor

REFERENCE

Bidlack, W.R. and Wang, W. 1998. Designing functional foods. In: *Modern Nutrition in Health and Disease*, Shils, M.D., Olson, J.S. and Shile, M., eds. Philadelphia, PA: Lea and Febiger.

Acknowledgements

The authors and editors wish to thank the Rehnborg Center for Nutrition Wellness, Nutrilite Division of Amway Corporation, for their support of the 1996 phytochemical conference, *Phytochemicals: Today's Knowledge—Tomorrow's Products* held in partnership with the College of Agriculture at the California State Polytechnic University, Pomona, November 18–19, 1996, which led to this publication. The attendees encouraged the publication of the material presented.

The editors thank Wei Wang, Ph.D., for her assistance in proof edits and indexing. We also thank the editorial staff and publishers at Technomic Publishing Co., Inc., for providing their quality efforts to bring this work to publication. Their support of this venture has made it possible for those who were unable to attend the conference to benefit from the international exchange of information.

List of Contributors

Frederick B. Askin
Division of Surgical Pathology
The Johns Hopkins Medical
  Institution
600 North Wolfe Street
Baltimore, MD 21287-6417

Ami Ben-Amotz
The National Institute of
  Oceanography
Israel Oceanographic and
  Limnological Research
Tel-Shikmona
P.O. Box 31080, Israel

Eric Block
Department of Chemistry
State University of New York at
  Albany
Albany, NY 12222

Cora J. Dillard
Department of Food Science and
  Technology
University of California
Davis, CA 95616

J. Bruce German
Department of Food Science and
  Technology
University of California
Davis, CA 95616

Ross L. Hood
The CACI Clinic
21 Copperleaf Place
Castle Hill, NSW 2154, Australia

Frederick Khachik
Department of Chemistry and
    Biochemistry
Joint Institute for Food Safety and
    Applied Nutrition
University of Maryland
College Park, MD 20742

Keith Lai
Division of Surgical Pathology
The Johns Hopkins Medical
    Institutions
600 North Wolfe Street
Baltimore, MD 21287-6417

Richard E. Litov
Director of Research and
    Development
NutraTec
3210 Arrowhead Drive
Evansville, IN 47720-2504

Stanley T. Omaye
Department of Nutrition and the
    Environmental Sciences and
    Health Graduate Program
University of Nevada
Reno, NV 89507

Inke Paetau
Phytochemical Laboratory USDA,
Beltsville Human Nutrition
    Research Center
Beltsville, MD 20892

Martin F. Stoner
Biological Sciences Department
California State Polytechnic
    University
Pomona, CA 91768

Christine A. Swanson
Nutritional Epidemiology Branch
    Epidemiology and Biostatistics
    Program, DCEG, NCI, NIH
Executive Plaza North, Suite 443
6130 Executive Boulevard
Bethesda, MD 20892-7374

Wendy S. White
Center for Designing Foods to
    Improve Nutrition
Department of Food Science and
    Human Nutrition
Iowa State University
Ames, IA 50011-1120

Peng Zhang
Department of Nutrition and the
    Environmental Sciences and
    Health Graduate Program
University of Nevada
Reno, NV 89507

CHAPTER 1

Vegetables, Fruits, and Cancer Risk: The Role of Phytochemicals

CHRISTINE A. SWANSON

INTRODUCTION

MORE than 50 years ago, animal studies provided clear evidence that diet played a role in the development and growth of tumors (Tannenbaumn, 1940). Interest in diet as a cause of human cancer is a relatively new phenomenon. Recent data (Willett, 1995) suggest that as much as one third (range 20–42%) of all cancer deaths in the United States are avoidable by dietary change. Thus, diet ranks second only to smoking as a leading contributor to cancer incidence and mortality. This chapter provides an overview of the relation of fruits and vegetables to cancer risk and considers the potential role of plant phytochemicals.

DIET AND CANCER RISK: FRUITS AND VEGETABLES

Based largely on geographical correlation studies, attention was focused initially on increased cancer risk associated with consumption of high-fat/low-fiber or "Westernized" diets. In recent years, the protective effect of fruits and vegetables has been emphasized. In 1991, Steinmetz and Potter (1991a) published the first comprehensive review of the relation between cancer risk and consumption of fruits and vegetables. They concluded that diets rich in fruits and vegetables were associated consistently with reduced

1

risk of a variety of tumors, especially cancers of the respiratory and gastrointestinal tract (e.g., lung, esophagus, stomach, and colon). In 1996, cancer epidemiologists updated current knowledge regarding the relation of diet and human cancers (Trichopoulos and Willett, 1996). A summary of the role of fruits and vegetables from that report and other reviews (Steinmetz and Potter, 1991a; Block et al., 1992; Stewart et al., 1996) is shown in Table 1. The literature indicates an almost universal protective role of fruits and vegetables. Overall, vegetables are more strongly related to risk reduction than are fruits. The benefit of diets rich in fruits and vegetables tends to be strongest for epithelial cancers of the respiratory and digestive tract. At present, none of the hormonally related cancers (i.e., breast, endometrium, ovary, and prostate) has been clearly related to intake of fruits or vegetables.

TABLE 1. Qualitative Relation of the Protective Effect of Vegetables and Fruits According to Cancer Type.

Cancer Type	Vegetables	Fruits
Mouth	+ +	+ +
Nasopharynx	+	+
Larynx	+ + +	+ + +
Lung	+ + +	+ +
Esophagus	+ + +	+ + +
Stomach	+ + +	+ + +
Liver	+	+
Pancreas	+	+
Kidney	+ +	+
Bladder	+ +	+ +
Colon	+ + +	+ +
Breast	+	+
Endometrium	+	+
Ovary	+	+
Cervix	+	+
Prostate	+	

+ + +Convincing evidence.
+ +Suggestive evidence.
+Weak but suggestive evidence.
Compiled from Steinmetz and Potter (1991a), Block et al. (1992), and Willett (1995).

EPIDEMIOLOGIC EVIDENCE: SELECTED EXAMPLES

LUNG CANCER

Of all the epidemiologic investigations of diet and cancer, lung cancer studies provide the most persuasive evidence of a protective effect of fruits and vegetables (Ziegler et al., 1996). Both case-control and prospective studies demonstrate that increased fruit and vegetable intake is associated with reduced risk of the disease. The association is observed in men and women; in smokers, ex-smokers, and nonsmokers; and for all histologic types of the disease. Across studies, lung cancer risk is increased anywhere from 30 to 100% [i.e., relative risks (RRs) of 1.3–2.0] among individuals in the lowest quintile of intake compared with those in the highest category. These risk estimates probably are underestimates of the true effect because dietary assessment methods appropriate for epidemiologic studies often result in significant exposure misclassification and attenuated risk estimates.

In numerous studies, yellow-orange and green leafy vegetables, particularly those rich in carotenoids, were found to be important. Further support for a protective effect of carotenoids was provided by studies that included biochemical determinations. Prospective studies of blood β-carotene concentration, also a good biomarker of fruit and vegetable intake, indicated that low levels were predictive of increased lung cancer incidence (Stahelin et al., 1984, Nomura et al., 1985). Although the active component(s) in fruits and vegetables was not identified, β-carotene was strongly suspected to be etiologically important (Peto et al., 1981).

Three large clinical trials were initiated to test the hypothesis that β-carotene supplementation at pharmacologic doses could reduce the incidence of lung cancer (Albanes and Heinonen, 1994; Omenn et al., 1996; Hennekens et al., 1996). The hypothesis was not borne out. On the contrary, supplementation with β-carotene was neither safe nor effective. Supplementation was associated with a significant excess of both lung cancer and total mortality in two high-risk populations comprised largely of male smokers (Albanes and Heinonen, 1994; Omenn et al., 1996). In the study of male health professionals (Hennekens et al., 1996), β-carotene had no effect.

Not surprisingly, the unexpected findings of the Alpha-Tocopherol, Beta-Carotene Study (ATBC) and the Beta-Carotene and Retinol Efficacy Trial (CARET) stimulated considerable discussion. Numerous explanations for the adverse effects of β-carotene have been postulated. Supplementation with pharmacological levels of β-carotene may have interfered with the absorption and utilization of other dietary constituents, including some bioactive oxycarotenoids (see Chapter 6). Adverse interactions with both alcohol consumption and smoking have been postulated. Subsequent analysis of the ATBC study, for example, showed that the adverse effect of β-carotene was largely restricted to drinkers (Albanes, personal communication). While β-carotene is generally thought of as an antioxidant, it also appears to have pro-oxidant properties under conditions of high oxygen tension, which would occur among smokers (Mayne et al., 1996).

We do not know why diets rich in fruits and vegetables are associated with reduced risk of lung cancer. In retrospect, we appreciate that there are hundreds of carotenoids in fruit and vegetables along with hundreds of other bioactive constituents. Despite the wealth of information provided by observational epidemiology, the results of intervention studies with β-carotene suggest that we have greatly underestimated the complexity of this particular diet-cancer association.

ESOPHAGEAL CANCER

Frequent consumption of fruits and vegetables has been shown repeatedly to be associated with decreased risk of esophageal cancer (Steinmetz and Potter, 1991a; Cheng and Day, 1996). In Western countries, alcohol and smoking are the predominant risk factors. However, after accounting for these variables, risk among individuals with the lowest intake of fruits and vegetables often is twice that of those having the highest intake (RR = 2.0). In a number of studies, fruit (particularly raw fruit and citrus) and dark yellow and green leafy vegetables were most protective. As is typical of many epidemiologic studies of diet and cancer, when comparing risk estimates associated with specific micronutrients (e.g., ascorbic acid) to those of the foods from which they are derived, the effects tend to be stronger for foods. This could indicate that other unidentified constituents in fruits and vegetables are important. Nutrients in food do not occur in isolation, perhaps it is the combination of nutrients and other dietary constituents of fruit and vegetables that is etiologically relevant (Block et al., 1992). Alternatively, food composition data may be inadequate (Riboli et al., 1996), resulting in misclassification and attenuation of risk estimates.

Esophageal cancer is not common in the United States but is endemic in some developing countries. Linxian county in Hunan Province, China, has one of the world's highest rates of cancer of the esophagus and of the esophageal/gastric cardia (Li, 1982). In 1985, four combinations of micronutrients were tested in a large clinical trial conducted in Linxian (Blot et al., 1993). After 5 years of supplementation, significantly lower total mortality occurred among individuals receiving β-carotene, vitamin E, and selenium daily. The reduction in mortality was due mainly to lower cancer rates, especially for gastric cancer. Risk of esophageal cancer was also reduced, but the finding was not statistically significant. This study was the first clinical trial to report a significant reduction in cancer risk from nutrient supplements. Perhaps the best explanation for the positive effect of supplementation was the prudent choice of a high-risk population with demonstrated nutrient deficiencies. There is no clear evidence that vitamin and mineral supplements are associated with reduced risk of esophageal or gastric cancer in Western populations.

PROSTATE CANCER

Prostate cancer has, until recently, received less attention than other major cancers. The role of dietary fat has been emphasized but remains unclear (Kolonel, 1996). Recently, the relation of intake of fruits, vegetables, and various carotenoids to prostate cancer was examined in a large prospective study of male health professionals (Giovannucci et al., 1995). Intake of β-carotene, α-carotene, lutein, and β-cryptoxanthin was not associated with risk of the disease. However, risk of prostate cancer was 21% lower (RR = 0.79, 95% confidence interval, CI, 0.64–0.99) among men in the highest quintile of lycopene intake compared with those in the lowest category. Sources of lycopene in the U.S. diet are relatively limited, tomatoes being the leading contributor. When intake of tomatoes, tomato sauce, tomato juice, and pizza were combined, risk of prostate cancer was further reduced: 35% lower (95% CI, 0.44-0.95) among those in the highest versus lowest quintile of intake (10 versus less than 1.5 servings per week). The results were not due to a correlation between lycopene and total vegetable intake and apparently were not confounded by other factors (e.g., socioeconomic status) related to risk of the disease. If confirmed, the results of this study are unique, suggesting a singularly protective effect of a specific food (tomatoes or tomato products) and possibly a specific dietary constituent, lycopene, which is not presently considered essential.

CLASSIFICATION OF FRUITS AND VEGETABLES

Classification schemes for fruits and vegetables (Smith et al., 1995), particularly those based solely on color or other physical characteristics (e.g., dark-green leafy vegetables), appear to be relatively arbitrary at first glance. In fact, many food groupings serve as useful surrogates for phytochemicals (e.g., carotenoids from yellow, orange, or dark-green leafy vegetables). Botanical families provide an alternative classification system. Some botanical groupings are useful proxies for phytochemicals that are essential nutrients (e.g., ascorbic acid from *Rutaceae* or citrus fruit). Other botanical families serve as indices of a number of plant constituents that are bioactive but not currently considered to be essential. These "nonessential" compounds may eventually prove to be relevant to human cancer risk. For example, animal studies demonstrate that organosulfur compounds found in allium vegetables inhibit the appearance of several types of tumors and also decrease tumor proliferation (Belman, 1983; Wargovich, 1987). Frequent consumption of allium vegetables (e.g., garlic, onion, leeks, and chives) has been associated with reduced risk of stomach cancer in humans (You et al., 1989). Unlike those for essential nutrients, databases for most of the nonessential phytochemicals of interest to cancer biologists and epidemiologists either do not exist or are very limited. We cannot use analytic epidemiologic studies to assign risk estimates to specific compounds without food composition data.

BIOLOGIC PLAUSIBILITY: A CONCEPTUAL FRAMEWORK

The pathway leading ultimately to the replication of neoplastic cells provides a mechanistic framework for discussing the role of phytochemicals in human carcinogenesis (Figure 1). To interact with and eventually damage DNA, a number of human carcinogens (some of them coming from plants themselves) must first be activated. Many mutagens never reach the target site because of elaborate enzyme systems forming a first line of defense. Phase I oxidative reactions of the cytochrome P450 superfamily, for example, render a number of potential mutagens more water soluble and thus more readily available for excretion. Other P450 enzymes, however, may activate carcinogens. Therefore, the chemopreventive effectiveness of a specific phytochemical will be determined, in large part, by its ability to favor detoxification reactions (Wattenberg, 1992; Wolf et al., 1996). Phase II enzymes, such as the glutathione-*S*-transferases (GSTs), facilitate conjugation and other reactions that inactivate or detoxify carcinogenic agents. If a carcinogen is not blocked or otherwise inactivated, it can interact with and damage DNA, a critical step in carcinogenesis. DNA repair enzymes efficiently remove most, but not all, of the lesions. A cell with DNA damage can give rise to a mutation when it divides. Finally, in the absence of nuclear control mechanisms regulating cel-

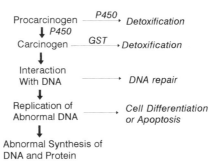

Figure 1 Molecular events leading to replication of neoplastic cells.

lular proliferation, neoplastic cells multiply. A number of compounds in fruits and vegetables are known to modulate these pathways.

Wattenberg (Wattenberg, 1992), has broadly classified phytochemicals as blocking or suppressing agents. Blocking agents prevent carcinogens from reaching or reacting with target sites (i.e., DNA). They work by a variety of mechanisms, including induction or detoxification systems (i.e., the phase I and phase II enzymes). Suppressing agents, on the other hand, prevent the progression of transformed cells that would otherwise become malignant. The number of suppressing agents identified in animal models is much smaller than the number of blocking agents, and their mechanisms of action are less well understood (Wattenberg, 1996).

Ames et al. (1993) have referred to biochemical pathways associated with carcinogen metabolism to give biological plausibility to the hypothesis that reduction in cancer risk associated with frequent intake of fruits and vegetables is, in large part, explained by antioxidants. Briefly, in the course of daily living, we are regularly exposed to mutagens in the form of reactive oxygen species (e.g., superoxide, hydrogen peroxide, and hydroxyl radicals). These compounds are the inevitable by-products of normal oxidative metabolism. Not surprisingly, we have evolved endogenous antioxidants (e.g., glutathione peroxidase) that are bolstered by exogenous defenses in the form of dietary antioxidants, including ascorbate, tocopherols, and some carotenoids. In addition to inactivating reactive oxygen species, antioxidants may also have a prominent role in regulating cell division (Ames et al., 1995).

BIOACTIVE COMPOUNDS IN FRUITS AND VEGETABLES

Until recently, nonessential bioactive compounds in fruits and vegetables have received less attention than essential nutrients. Selected nonessential phytochemicals, mostly blocking agents, from plant sources are listed in Table 2. Detailed descriptions of these and other phytochemicals are provided in review articles (Steinmetz and Potter, 1991b; Wattenberg, 1992, Kitts, 1994).

While we tend to think of phytochemicals as minor dietary constituents, phenolic compounds are widely distributed in fruit and vegetables and may constitute as much as 10 to 20% of the dry weight of the diet of a strict vegan (Steinmetz and Potter, 1991b), roughly 40 to 80 grams. Some phenolics trap nitrates and prevent the formation of mutagenic *N*-nitroso compounds (Stich, 1984). Exposure to *N*-nitroso compounds has been linked to cancers of the nasopharynx, esophagus, and stomach (Bartsch and Montesano, 1984). Other phenolics have been shown to induce detoxification systems, especially phase II conjugation reactions (Steinmetz and Potter, 1991b). Phenolics have been extracted from plant products other than fruits and vegetables. Tea polyphenols, for example, have been shown to reduce tumors in experimental animals

TABLE 2. Selected Nonessential but Bioactive Compounds
in Edible Plants.

Compounds	Food Source
Phenolics	fruits, vegetables, soybeans, cereals, tea, coffee
elagic acid	vegetables
curcumin	curry spice
coumarin	vegetables, citrus fruits
Organosulfides	allium foods (garlic, onion, leeks)
diallyl sulfide	
Glucosinolates	cruciferous vegetables
isothiocyanates	
Indols	cruciferous vegetables
indole-3-carbinol	
Flavones	fruits, vegetables
quercetin	berries, tomatoes, potatoes
tangeretin	citrus fruit
rutin	citrus fruit
Isoflavones	flax seed, lentils, soybeans
genistein	soy
daidzein	soy

Compiled from Steinmetz and Potter (1991b), Wattenberg (1992), and Kitts (1994).

(Yang and Wang, 1993). The anticancer effect of tea phytochemicals may be dose dependent. At high concentrations, tea extracts can effectively block endogenous formation of N-nitroso compounds, whereas nitrosation reactions may be enhanced at low levels (Yang and Wang, 1993). Green tea, rich in polyphenolics, has been associated with decreased risk of esophageal cancer in China (Gao et al., 1994).

Animal studies provide compelling evidence that compounds in allium foods are associated with reduced risk of a variety of tumors (Steinmetz and Potter, 1991b). This class of foods is rich in organosulfides, of which diallyl sulfide has been studied extensively. Allium foods may block the formation of N-nitroso compounds by inhibiting the bacterial conversion of nitrate to nitrite (Steinmetz and Potter, 1991b). They also have been shown to induce detoxification systems (Sparnins et al., 1986). In vitro studies indicated that onion extract decreased tumor proliferation (Niukian et al., 1987).

Cruciferous vegetables are rich in several classes of compounds with clear anticancer activity in experimental animals. Almost two dozen glucosinolates have been identified in cruciferous vegetables (Steinmetz and Potter, 1991b). Among them, isothiocyanates have been shown to induce phase II xenobiotic-metabolizing enzymes including GST. Studies of indoles in cruciferous vegetables demonstrate the complexity of enzyme systems regulating the activation of carcinogens (Steinmetz and Potter, 1991b). Indole-3-carbinol, for example, has been shown to increase mixed function oxidase activity. Subsequently, induction of this system can either activate or detoxify carcinogenic compounds.

Flavonoids, which are widely distributed in fruit and vegetables, also have produced mixed results with respect to cancer-related effects. The protective effect of some flavonoids may be explained by their antioxidant activity. Increased incidence of tumors by other flavonoids may be related to a balance of phase I enzymes favoring carcinogen activation rather than detoxification (Friedman and Smith, 1984). Flavonoids represent one of the few non-nutrient phytochemicals for which food values for specific compounds have been published (Herrmann, 1976).

Isoflavones, which are closely related in structure to estrogenic steroids, have generated considerable research attention in the last decade. Some researchers (Adlercreutz, 1990; Messina et al., 1994) have hypothesized that plant estrogens, including genistein and daidzein, may block more potent endogenous estrogens, presumably reducing the risk of estrogen-dependent tumors, particularly those of the breast and endometrium. Some isoflavones may exert their protective effect by inhibiting protein kinases involved in signal transduction (Akiyama et al., 1991).

FRUIT AND VEGETABLE INTAKE IN THE UNITED STATES

Most Americans do not eat enough fruits and vegetables. *Dietary Guidelines for Americans* issued by the Departments of Agriculture and Health and Human Services state that Americans should eat at least 2 servings of fruits and 3 servings of vegetables each day. Based on single 24-hour recalls, only 9% of adults who participated in the second National Health and Nutrition Examination Survey (NHANES II, 1976–1980) met these guidelines on any given day. Approximately 45% of the population consumed no fruits or fruit juice, and 22% had no servings of vegetables on the recall day (Patterson and Block, 1988).

To improve fruit and vegetable intake in the United States, the National Cancer Institute of the National Institutes of Health initiated the Five a Day Program (Havas, 1994). The purpose of this effort is to promote consumption of five or more servings of fruits and vegetables daily. The Five a Day Baseline Survey of almost 3,000 adults indicated that the median intake of fruits and vegetables was 3.4 servings daily, again far short of the recommendation (Subar et al., 1995). Only 23% of the population regularly consumed five or more servings daily.

SUMMARY

Both animal and human studies provide clear evidence that diets rich in fruits and vegetables are associated with reduced risk of a variety of cancers. The beneficial effect of vegetables is clear. The effect of fruits is somewhat less persuasive. In general, the effect of whole foods is more pronounced than that of the individual micronutrients or other bioactive phytochemicals they supply. Sorting out the possible protective components in fruits and vegetables has been extremely difficult, but numerous "nonessential" dietary plant constituents clearly have anticarcinogenic potential. The role of these compounds in human cancer, however, is not established. Epidemiologic data for these nonessential but bioactive plant constituents are sparse, in large part, because food composition data are lacking. While there is uncertainty as to which dietary phytochemicals or combinations of plant constituents confer protection, a recommendation for Americans to increase their consumption of fruits and vegetables can be made without reservation.

REFERENCES

Adlercreutz, H. Western diet and Western diseases: some hormonal and biochemical mechanisms and associations. *Scand. J. Clin. Lab. Invest.* 50(Suppl 201):3–23, 1990.

Akiyama, T., Ogawara, H. Use and specificity of genistein as inhibitor of protein-tyrosine kinases. *Method. Enzymol.* 201:362–370, 1991.

Albanes, D. A., Heinonen, O. P., The ATBC Cancer Prevention Study Group. The effect of vitamin E and beta carotene on the incidence of lung cancer and other cancers in male smokers. *N. Engl. J. Med.* 330:1029–1035, 1994.

Ames, B. N., Shigenaga, M. K., Hagen, T. M. Oxidants, antioxidants, and the degenerative diseases of aging. *Proc. Natl. Acad. Sci. USA* 90:7915–7922, 1993.

Ames, B. N., Gold, L. S., Willett, W. C. The causes and prevention of cancer. *Proc. Natl. Acad. Sci. USA*, 92:5258–5265, 1995.

Bartsch, H., Montesano, R. Relevance of nitrosamines to human cancer. *Carcinogenesis* 5(11):1381–1393, 1984.

Belman, S. Onion and garlic oils inhibit tumor promotion. *Carcinogenesis* 4:1063–1065, 1983.

Block, G., Patterson, B., Subar, A. Fruit, vegetables, and cancer prevention: a review of the epidemiological evidence. *Nutr. Cancer* 18:1–29, 1992.

Blot, W. J., Li, J. Y., Taylor, P. R., Guo, W., Dawsey, S., Wang, G. Q., Yang, C. S., Zheng, S. F., Gail, M., Li, G. Y., Yu, Y., Liu, B. Q., Tangrea, J., Sun, Y. H., Liu, F., Fusheng, L., Fraumeni, J. F. Jr., Zhang, Y. H., Li, B. Nutrition intervention trials in Linxian, China: supplementation with specific vitamin/mineral combinations, cancer incidence, and disease-specific mortality in the general population, *J. Natl. Cancer Inst.* 85:1483–92, 1993.

Cheng, K. K., Day, N. E. Nutrition and esophageal cancer. *Cancer Causes and Control* 7:19–32, 1996.

Chug-Ahuja, J. K., Holden, J. M., Forman, M., Mangels, A. R., Beecher, G. R., Lanza, E. The development and application of a carotenoid database for fruits, vegetables, and selected multicomponent foods. *J. Am. Diet. Assoc.* 93:318–323, 1993.

Friedman, M., Smith, G. A. Factors which facilitate inactivation of quercetin mutagenicity. *Adv. Exp. Med.* 177:527–44, 1984.

Gao, Y. T., McLaughlin, J. K., Blot, W. J., et al. Reduced risk of esophageal cancer associated with green tea consumption. *J. Natl. Cancer Inst.* 86:855–8, 1994.

Giovannucci, E., Ascherio, A., Rimm, E. B., Stampfer, M., Colditz, G. A., Willett, W. C. Intake of carotenoids and retinol in relation to risk of prostate cancer. *J. Natl. Cancer Inst.* 87:1767–1776, 1995.

Havas, S., Heimendinger, J., Damron, D., Nicklas, T.A., Cowan, A., Beresford, S.A., Sorenson, G., Buller, D., Bishop, D., Baranowski, T., Reynolds, K. 5 A Day for Better Health—Nine Community Research Projects to Increase fruit and vegetable consumption. *Public Health Rep* (QJA); 110:68–79, 1995.

Hennekens, C. H., Buring, J. E., Manson, J. E., Stampfer, M., Rosner, B., Cook, N. R., Belanger, C., LaMotte, F., Gaziano, J. M., Ridker, P. M., Willett, W., Peto, R. Lack of effect of long-term supplementation with beta carotene on the incidence of malignant neoplasms and cardiovascular disease. *N. Engl. J. Med.* 334:1145–1149, 1996.

Herrmann, K. Flavonols and flavones in food plants: a review. *J. Food Technol.* 11:433–438, 1976.

Kitts, D. D. Bioactive substances in food: identification and potential uses. *Can. J. Physiol. Pharmacol.* 72:423–434, 1994.

Kolonel, L. N. Nutrition and prostate cancer. *Cancer Causes and Control* 7:83–94, 1996.

Li, J. Y. Epidemiology of esophageal cancer in China. *Monogr. Natl. Cancer Inst.* 62:113–120, 1982.

Mayne, S. T., Handelman, G. J., Beecher, G. Beta-carotene and lung cancer promotion in heavy smokers—a plausible relationship. *J. Natl. Cancer Inst.*, 88(21):1513–1515, 1996.

Messina, M. J., Persky, V., Setchell, K. D. R, Barnes, S. Soy intake and cancer risk: a review of the in vitro and in vivo data. *Nutr. Cancer* 21:113–131, 1994.

Niukian, K., Schwartz, J., Shklar, G. In vitro inhibitory effect of onion extract on hamster buccal pouch carcinogenesis. *Nutr. Cancer* 10:137–144, 1987.

Nomura, A. M. Y., Stemmermann, G. N., Helibrun, L. K., Salkeld, R. M., Vuilleumier, J. P. Serum vitamin levels and the risk of cancer of specific sites in men of Japanese ancestry in Hawaii. *Cancer Res.* 45:2369–2372, 1985.

Omenn, G. S. What accounts for the association of vegetables and fruits with lower incidence of cancers and coronary heart disease? *Ann. Epidemiol.* 5:333–335, 1995.

Omenn, G. S., Goodman, G. E., Thornquist, M. D., Balmes, J., Cullen, M. R., Glass, A., Keogh, J. P., Meyskens, F. L. Jr., Valanis, B., Williams, J. H. Jr., Barnhart, S., Hammar, S. Effects of a combination of beta carotene and vitamin A on lung cancer and cardiovascular disease. *N. Engl. J. Med.* 334:1150–1155, 1996.

Patterson, B. H., Block, G. Food choices and the cancer guidelines. *Am. J. Public Health* 78:282–286, 1988.

Peto, R., Doll, R., Buckley, J. D., Sporn, M. B. Can dietary beta-carotene materially reduce human cancer rates? *Nature* 290:201–208, 1981.

Riboli, E., Slimani, N., Kaaks, R. Identifiability of food components for cancer chemoprevention. In: *Principles of Chemoprevention.* Stewart, D., McGregor, D., Kleihues, P. (eds.) Lyon France: IARC Scientific Publications, No. 139; pp. 165–173, 1996.

Smith, S. A., Campbell, D. R., Elmer, P., Martini, M. C., Slavin, J. L., Potter, J. D. The University of Minnesota Cancer Prevention Research Unit vegetable and fruit classification scheme (United States). *Cancer Causes Control* 6:292–302, 1995.

Sparnins, V. L., Mott, A. W., Barany, G., Wattenberg, L. W. Effect of allyl methyl trisulfide on glutathione *S*-transferase activity and BP-induced neoplasia in the mouse. *Nutr. Cancer* 8:211–215, 1986.

Stahelin, H. B., Rosel, F., Buess, E., Brubacher, G. Cancer, vitamins, and plasma lipids: prospective Basel study. *J. Natl. Cancer Inst.* 73:1463–1468, 1984.

Steinmetz, K. A., Potter, J. D. Vegetables, fruit, and cancer. I. Epidemiology. *Cancer Causes Control* 2:325–357, 1991a.

Steinmetz, K. A., Potter, J. D. Vegetables, fruit, and cancer. II. Mechanisms. *Cancer Causes Control* 2:427–442, 1991b.

Stewart, D., McGregor, D., Kleihues, P. (eds.). *Principles of Chemoprevention.* Lyon France: IARC Scientific Publications, No. 139, 1996.

Stich, H. F., Rosin, M. P. Naturally occurring phenolics as antimutagenic and anticarcinogenic agents. *Adv. Exp. Med.* 177:1–29, 1984.

Subar, A. F., Heimendinger, J., Patterson, B. H., Krebs-Smith, S. M., Pivonka, E., Kessler, R. Fruit and vegetables intake in the United States: the baseline survey of the Five A Day For Better Health Program. *Am. J. Health Promot.* 9(5):352–360, 1995.

Tannenbaum, A. The initiation and growth of tumors: introduction. I. Effects of underfeeding. *Am. J. Cancer* 38:335–350, 1940.

Trichopoulos, D., Willett, W. C. (eds.). Nutrition and cancer. *Cancer Causes Control* 7:1, 1996.

Wargovich, M. J. Diallyl sulfide, a flavor component in garlic (*Allium sativum*), inhibits dimethylhydrazine-induced colon cancer. *Carcinogenesis* 8:487–489, 1987.

Wattenberg, L. W. Inhibition of carcinogenesis by minor dietary constituents, *Cancer Res.* 52:2085s–92s, 1992.

Wattenberg, L. W. Inhibition of tumorigenesis in animals. *IARC Sci. Publ.* 139:151–158, 1996.

Willett, W. C. Diet, nutrition and avoidable cancer. *Environ. Hlth. Perspect.* 103s:165–70, 1995.

Willett, W. C., Trichopoulos, D. Nutrition and cancer. *Cancer Causes Control* 7(1):178–180, 1996.

Wolf, C. R., Mahmood, A., Henderson, C. J., McLeod, R., Manson, M. M., Neal, G. E., Hayes, J. D. Modulation of the cytochrome P450 system as a mechanism of chemoprevention. *IARC Sci. Publ.* 139:165–73, 1996.

Yang, C. S., Wang, Z. Y. Tea and cancer. *J. Natl. Cancer Inst.* 85:1038–1049, 1993.

You, W. C., Blot, W. J., Chang, Y. S., Ershow, A., Yang, Z. T., An, Q., Henderson, B. E., Fraumeni, J. F. Jr., Wang, T. G. Allium vegetables and reduced risk of stomach cancer. *J. Natl. Cancer Inst.* 81:162–164, 1989.

Zeigler, R. G., Taylor-Mayne, S., Swanson, C. A. Nutrition and lung cancer. *Cancer Causes Control* 7:157–177, 1996.

CHAPTER 2

Phytochemicals and Targets of Chronic Disease

J. BRUCE GERMAN
CORA J. DILLARD

INTRODUCTION

CONCOMITANT with recognition of the significance of oxidative stress in the etiology of chronic diseases such as cancer and autoimmunity, cardiovascular disease, postischemia injury, trauma, and aging has been the recognition that certain antioxidant nutrients may chemically retard these processes. This recognition has led to the hypothesis that ingestion of antioxidant plant components may prevent or ameliorate these disease processes. Prime among these antioxidant components in plants are the polyphenolic constituents generally classified as phytochemicals. An understanding of the complex nature of the chemical process of oxidation and the biochemistry involved in the absorption, metabolism, and molecular functions of phytochemicals is requisite to the understanding of how they may affect health status.

Oxidation is the chemical process by which oxygen adds to and withdraws energy from reduced carbon-based molecules. The paradox is that this process of free radical oxidative reactions is both deleterious and life sustaining by being coupled to electron transport in the mitochondria of living cells. A complex spectrum of biochemical systems that either utilize or detoxify products of free radical chain reactions has gradually developed over the course of the evolution of organisms living in an oxygen atmosphere. Scientists are only now recognizing and beginning to understand the interactions between these systems that generate and utilize oxidants. This understanding will lead to knowledge of where in the chemical process normal reactions

13

begin to deviate toward pathological reactions and where antioxidant phytochemicals can provide protection in living systems against some disease processes.

Oxidant exposure and antioxidant depletion are general phenomena that together are described as "oxidative stress." Both chronic and natural and both acute and catastrophic events comprise oxidative stress (Halliwell and Gutteridge, 1989). Oxidative stress induced by chemical toxicity or iron overload (Halliwell, 1991; Gutteridge, 1994; Stohs and Bagchi, 1995) results in gross tissue damage, while normal metabolism, such as macrophage release activated oxygen to kill invading pathogens, can result in mild, yet chronic damage (Ames et al., 1993). Thus, a defensive strategy that generates chemically toxic oxygen radicals is balanced by protection against the threat of systemic infection. The chronic inflammation of autoimmune disorders, however, creates pathological changes (Keen et al., 1991; Bashir et al., 1993). This seemingly two-edged sword underscores both the protective power of oxidative chemistry and the risks associated with these naturally evolved biological reactions.

The balance between the benefits and risks of free radical oxidative chemistry is hypothesized to be shifted by changing the predominance of pro-oxidative processes and antioxygenic protection. The hypothesis has been tested by epidemiological studies of various populations wherein disease incidence has been related to genetic and environmental variables. Such studies suggest that a disparity between antioxidant protectors and oxidative stress may lead to repetitive damage to sensitive biological tissues. The oxidant-antioxidant imbalance appears to promote buildup of potentially toxic oxidant response and repair system by-products, which appears to lead to the deterioration of function.

The concept of oxidant-antioxidant balance has profound implications for studies of the etiology of chronic disease. Recently, nonessential dietary components have been recognized as potentially participating in this balance. For example, wine polyphenolic phytochemicals were proposed to retard low-density lipoprotein (LDL) modification (Frankel et al., 1993), an event considered causal to atherosclerosis. This review emphasizes the multistage nature of oxidative chemistry within biological systems. The multiple mechanisms of pathogenesis and the pervasive nature of free radical oxidative events are thought to be the basis for the benefits of dietary polyphenolic phytochemicals.

CHEMISTRY OF OXIDATION

Thermodynamic equilibrium strongly favors the net oxidation of reduced, carbon-based biomolecules. The kinetic stability of all biological molecules in an oxygen-rich atmosphere results from the unique spin state of the unpaired electrons in ground-state molecular (triplet) oxygen in the atmosphere. This property renders atmospheric oxygen relatively inert to reduced, carbon-based biomolecules. Hence, reactions between oxygen and protein, lipids, polynucleotides, and carbohydrates proceed at vanishingly slow rates unless they are catalyzed. However, once a free radical chain reaction is initiated, the free radicals generated rapidly propagate and interact directly with various targets and also yield hydroperoxides. The hyperperoxides are readily attacked by reduced metals, leading to a host of decomposition products. Some of these products cause further damage, and some, formed through self-propagating reactions, are themselves free radicals; thus, oxidation is reinitiated. A large volume of literature points to several key participants in the reaction course (Frankel, 1991; Porter et al., 1995).

Initiators of oxidation eliminate the reactive impediments imposed by the spin restrictions of ground-state oxygen by converting stable organic molecules, RH, to free radical-containing molecules, $R^•$. Oxygen reacts readily with such species to form the peroxy radical, $ROO^•$. Initiators of lipid oxidation are relatively ubiquitous, primarily single-electron oxidants, and include trace metals, hydroperoxide cleavage products, and light. A risk of biological systems that use polyunsaturated fatty acids (PUFAs) is that these molecules are oxidized by the $ROO^•$ species to yield another free radical, $R^•$, and a lipid hydroperoxide, ROOH. This effectively sets up a self-propagating free radical chain reaction, $R^• + O_2 \rightarrow ROO^• \rightarrow ROOH + R^•$, that can lead to the complete consumption of PUFA in a free radical chain reaction (Kanner et al., 1987). The ability of the peroxy radical to act as an initiating, single-electron oxidant drives the destructive and self-perpetuating reaction of PUFA oxidation.

Lipid hydroperoxides formed during lipid oxidation decompose to short-chain aldehydes, ketones, and alcohols (Frankel, 1991). These products, as well as radicals, compromise health a number of ways: (1) The direct oxidation of susceptible molecules can result in loss of function. For example, oxidation of membrane lipids alters membrane integrity and promotes red blood cell fragility and membrane leakage. The oxidation of proteins results in loss of enzyme catalytic activity and/or regulation. (2) Reaction of some of these products leads to adduct formation with loss of native functions. The oxidative modification of the apolipoprotein B molecule on LDL prevents uptake by the LDL receptor and stimulates uptake by the scavenger receptor. (3) Oxidation can cleave DNA and cause point, frameshift, and deletion mutations and base damage. This oxidative cleavage impairs or destroys normal functionality. (4) Oxidative reactions can liberate signal molecules or analogs that elicit inappropriate responses such as the activation of platelet aggregation, promotion of cell proliferation, and down-regulation of vascular relaxation by leukotoxin and eicosanoid analogs.

The susceptibility and overall rate of oxidation of a lipid molecule is related to the number of double bonds on the fatty acids. The rate of oxidation is determined by the ease of hydrogen abstraction. An increase in the number of double bonds increases the oxidation rate; for example, the fatty acids 18:1, 18:2, 18:3, and 20:4 have relative oxidation rates of 1, 50, 100, and 200. These relative rates of oxidation as a function of number of double bonds may be important to rates of deterioration of biological molecules *in vivo.* Diets high in PUFA require more antioxidant nutrients to prevent oxidation and rancidity (Fritsche and Johnston, 1988). Consumption by animals of diets with high amounts of PUFA appears to increase the antioxidant requirement to prevent tissue damage (Muggli, 1989). The molecular basis for this increased requirement is not known. Frequently, reports of increased *in vivo* oxidative damage are based on crude measures of lipid oxidation such as thiobarbituric acid-reacting substances. The thiobarbituric acid assay does not distinguish oxidation among different dietary fats as it responds differently to the same amount of oxidation in PUFA with different numbers of double bonds. A diet enriched in highly unsaturated fatty acids would appear to increase the tendency to oxidation and increase the incidence of oxidation-associated chronic degenerative diseases, however, this has not been observed. In fact, studies have shown that replacement of diets high in saturated fat with highly unsaturated fat diets frequently reduces atherosclerosis, thrombosis, and other chronic diseases (Keen et al., 1991). Thus, when considering the myriad effects of dietary fats on tissue oxidation, it is critical to understand all of the various biochemical and metabolic consequences as well. Since oxidative processes are initiated through a variety of chemical and enzymatic reactions (Kanner et al., 1987), the inhibition of oxidative biochemical and metabolic pathways by phenolic phytochemicals may significantly alter their net contribution to oxidative damage to tissues.

ANTIOXIDANT CONTROL OF OXIDATION

The chemistry of free radical oxidations is multistage and complex. Oxidation is not a single catastrophic event. There is no single initiating oxidant that generates all free radicals; there are a great many sources of single-electron oxidants. Similarly, there is no single reactive product of oxidation; there are classes of products, many of which are both selectively and broadly damaging. Free radicals and their products react with virtually all biological molecules, and there is no single defense against all targets of oxidative damage. Thus, organisms have evolved a spectrum of mechanisms to prevent or respond to oxidative stresses and free radicals and their products at one or more of the many steps of oxidation.

The potential health effects of phytochemicals as antioxidant protectors must be considered in the context of the overall response of living organisms to oxidation. As illustrated by specific examples below, many complex biochemical pathways have evolved to respond to oxidation.

(1) Prevention of oxidant formation: Mitochondria and peroxisomes are specialized cellular organelles that contain the generation and transfer of electrons and that dispose of toxic intermediates and products of metabolism. Transferrin and ceruloplasmin are plasma proteins that actively sequester iron and copper ions, which are capable of initiating oxidation.

(2) Scavenging of activated oxidants: Primary chain reaction-breaking antioxidants include α-tocopherol, ubiquinone, ascorbate, uric acid, polyphenolics, various flavonoids and their polymers, amino acids, and protein thiols. The metabolic and chemical consumption of amino acids and protein thiols during oxidation implies that they can and do act as antioxidants. The extent of their participation in oxidative protection thus represents an ongoing functional and energetic investment in antioxidation by protein metabolism (Levine et al., 1996).

(3) Reduction of reactive intermediates: Higher animals possess enzyme systems that scavenge active oxygen and detoxify reactive intermediates, including catalase, a scavenger of hydrogen peroxide, glutathione peroxidase, a remover of hydroperoxides, and superoxide dismutase, which reduces superoxide anion. Natural food constituents with antioxidant activity can also act as free radical quenchers, antioxidants, and/or protectors/regenerators of other antioxidants. Synergistic (Buettner, 1993) and antagonistic effects among mixtures of antioxidant compounds are possible based on the nature of redox couples formed by the actual redox potential of these compounds present in tissues. As such, the final participation in a redox chain would vary among different tissues and according to dietary intakes. Phenolic antioxidants stabilize some enzymes, enhance some activities, and inhibit others.

(4) Induction of repair systems: Proteases, lipases, RNAases, etc. constantly turn over cellular constituents, and degradative enzymes often have higher affinities for modified molecules. Substrate affinities of synthetic enzymes discriminate against oxidative forms of lipids, proteins, and nucleotides. This discriminatory process removes damaged molecules from the cell. Oxidation of protein transcription factors such as NF-kappa B (Suzuki et al., 1995) changes their binding properties. Oxidized transcription factors bind to oxygen-response elements of promoters for selected genes. This is a means for the cell to detect oxidant levels and to induce gene families (e.g., genes encoding peroxidases) to properly respond to the stress.

(5) Apoptosis: Cells that are unrepairable are removed by the process of "programmed cell death" or apoptosis. This process, in some cases involving activation of oxygen-response elements, is used by higher organisms to

selectively reshape tissues during normal growth and development and when undergoing subnecrotic stress. During apoptosis, cells are disassembled in a highly regulated and coordinated fashion. Apoptosis also limits damage as it prevents the release of reactive and toxic compounds from lysosomes, peroxisomes, etc.

The complexity and interdependence of the systems described above indicate that oxidative stress could increase requirements for not only direct antioxidants but also those nutrients essential for proper up-regulation of oxidant defense and repair mechanisms.

ANTIOXIDANTS

Antioxidant is a broad classification for molecules that may act prior to, or during, a free radical chain reaction at initiation, propagation, termination, decomposition, or subsequent reaction of oxidation products at sensitive targets. Antioxygenic compounds can participate in several of the protective strategies described for higher animals. Differences in point of activity are not trivial and influence the efficacy of a given compound to act as a net antioxidant or protectant. Different molecular behavior can also affect the impact of oxidation, and its inhibition, on biological function and damage. The alkyl radical, $R^\bullet$, is too reactive in an oxygen-rich environment for many competing species to successfully re-reduce $R^\bullet$ to RH before oxygen adds to form the peroxy radical, $ROO^\bullet$. At this point, however, $ROO^\bullet$ is a relatively stable free radical that reacts comparatively slowly with targets such as PUFA. This is the most widely accepted point of action for free radical-scavenging antioxidants such as the phenolic tocopherol. Tocopherol can reduce $ROO^\bullet$ to ROOH with such ease that tocopherol is competitive with biologically sensitive targets such as unsaturated lipids, RH, even at 10,000-fold lower concentration. The tocopheroxyl radical, $A^\bullet$, is, in general, a poor oxidant and reacts significantly more slowly than $ROO^\bullet$. Conversion of $ROO^\bullet$ to ROOH and formation of $A^\bullet$ effectively imparts a kinetic hindrance on the propagating chain reaction. The tocopheroxyl radical can be re-reduced by reductants such as ascorbate, dimerized with another radical, or further oxidized to a quinone. These free radical-scavenging functions of tocopherol are well documented (Buettner, 1993).

Tocopherols act as a vitamin, although the subtle variations in the structures of different tocopherols result in different effective activities as vitamin E. The free radical-scavenging properties toward cell membrane peroxy radicals are believed to be the basis for the essentiality of the tocopherols as vitamin E. Similarly, the lack of free radical scavenging leads to the pathologies associated with the deficiency of tocopherols. That the basis for the essentiality of tocopherols lies in their ability to prevent oxidative damage raises an important nutritional question, "Are these actions also provided by nonessential polyphenolics present in plants and foods derived from them?" Many phytochemicals have been implicated as being capable of interfering with and inhibiting free radical chain reactions of lipids. Plant phenolics inhibit lipid hydroperoxide formation catalyzed by metals, radiation, and heme compounds (Buettner, 1993; Hanasaki et al., 1994) and also scavenge peroxy, alkoxy, and hydroxy radicals and singlet oxygen (Laughton et al., 1991; Tournaire et al., 1993; Hanasaki et al., 1994). Tocopherol in oxidizing lipid systems are spared by flavonoids (Jessup et al., 1990; Terao et al., 1994). If α-tocopherol is an essential antioxidant that acts where no other compound can, the sparing effect of nonessential antioxidants may be one of their most important actions.

Antioxidant activity is not limited to prevention of hydroperoxide formation. Hydroperoxides are not damaging to foods or biological molecules, but their presence is an indication that oxidation has occurred. Although hydroperoxides are not directly damaging, their decomposition by reduced metals generates the reactive hydroxyl radical, $HO^\bullet$, or the alkoxyl radical, $RO^\bullet$. These strongly electrophilic oxidants react with and oxidize virtually all biological macromolecules. The alkoxy radical typically fragments the parent lipid molecule and liberates electrophilic aldehydes, hydrocarbons, ketones, and alcohols. Both the highly reactive hydroxy and alkoxy radicals and the electrophilic aldehydes liberated with their reduction react readily with polypeptides (proteins) and polynucleotides (DNA). Thus, additional antioxidant actions include preventing hydroperoxide decomposition, reducing alkoxyl radicals, or scavenging the electrophilic aldehydes. The efficacy of different antioxidants varies during this phase of the oxidation process. Even tocopherol isomers differ with respect to their ability to prevent decomposition of hydroperoxides (Huang et al., 1994, 1995). Plant phenolics vary in their ability to interrupt a free radical chain reaction with differences detectable among different lipid systems, oxidation initiators, and other antioxygenic components.

A BALANCE BETWEEN PRO-OXIDANTS AND ANTIOXIDANTS

Organisms need to cascade and amplify chemical signals to develop appropriate responses to many stressors, including oxidants. Many of these signaling systems are oxidant-generating pathways, such as the enzymatic systems that oxidize specific PUFA moieties to form potent signaling molecules called oxylipins. These molecules, including prostaglandins, leukotrienes, etc., signal a state of stress to adjacent or responsive cells. Enzymatically produced oxidized lipids act on higher-order brain functions such as pain and even sleep. Oxidation of some protein transcription factors allows binding to "oxidant-response elements" within DNA and directly affects its transcription. This cascading proliferation of oxidized molecules, which accomplishes the tasks of intracellular and multicellular signaling, also places a burden on oxidant defense systems. These chemical signals clearly coevolved with the increasingly sophisticated and necessary oxidant repair systems. Perhaps the presence of these oxidant defenses allowed the proliferation of nonlethal uses of oxidants in higher organisms.

Oxidation is initiated in cells, tissues, and fluids by a host of chemical and protein (enzyme) factors. Oxidation events, such as reactions involving Fenton chemistry, mitochondrial electron transport, respiratory bursts, oxygenating enzymes, reductive cleavage of peroxides, and xenobiotic metabolism, are initiated by organisms because they are either essential for, or beneficial to, the success of the organism. As delineated, the pathogenic potential for these reactions can be considerable.

(1) Fenton chemistry: The univalent reduction of hydroperoxides by transition metals, especially ferrous and cuprous salts, yields an unpaired electron as either the alkoxyl or hydroxyl radical. The most destructive free radical-initiating system is probably the presence of free metals and oxidized lipids in atherosclerotic plaque. This event is strong evidence that oxidation of tissues and lipoproteins contributes to lesion progression.

(2) Mitochondrial electron transport: Mitochondria transfer billions of electrons per day into oxygen in single-electron steps that eventually convert available hydrocarbons into carbon dioxide and water. Even with transfer equal to 99.999% of total electrons transferred, leakage from this system is a significant source of reactive oxygen species. Release of free radicals has been estimated to be in excess of 10^6 free radical species per day. This imperfection of mitochondrial oxidative coupling is the basis of the mitochondrial damage theory of aging.

(3) Respiratory burst: The immune system has evolved a "leaky" oxygen-reducing system to combat pathogenic organisms. Stimulation of the terminal oxidase in phagocytes produces a burst of oxygen consumption that is stoichiometrically converted to superoxide. Superoxide kills cells in the immediate vicinity of the oxidant burst but is nonspecific and can damage host cells as well as invading pathogens. The animate immune response is a good example of the overall cost-benefit equation that is inherent to animal stress responses.

(4) Oxygenating enzymes: Many, and perhaps all, cells respond to external stimuli by liberating PUFA from their membranes. Arachidonic acid is particularly labile, and it initiates a signal cascade that depends upon its free radical-catalyzed, enzymatic oxygenation to hydroperoxide deriva-

tives broadly termed eicosanoids. These stereospecific molecules are local signals that activate that cell and its immediate neighbors, thus acting as part of the cellular stress response system in higher organisms. Chronic or overactivation of this system constitutes an oxidative stress, referred to as "peroxide tone," that produces inflammation and activates additional response systems.

(5) Reductive cleavage of peroxides: A general class of enzymes called peroxidases eliminates peroxides as a general detoxifying mechanism. Certain of the peroxidases catalyze oxidant production as a result of this reaction. Peroxidase oxidant production generally accompanies a response to pathogen invasion and so are considered additional elements in the killing mechanisms of immune cells. Chronic or overactivation of this system constitutes an independent oxidative stress.

(6) Xenobiotic metabolism: The primary tissue involved in the conversion of toxic chemicals into excretable compounds is the liver. This conversion is affected by inducible enzymes that are typically oxidases. As a result, toxin metabolism can and does produce free radical species. The poisonous properties of a host of pesticides, for example, are now recognized to result from the secondary by-products produced during the metabolism of xenobiotics.

The above-described use of oxidation by cells and tissues is a clear risk-benefit relationship. Such risks are acceptable, at least in the short term, because oxidation provides a net benefit. The long-term consequences may only be relevant to aging organisms and poorly defended tissues, both of which are found in humans.

The current view is that many chronic diseases are a result of unprotected or aberrant oxidation. Oxidant damage that accrues over a lifetime can greatly influence the health of the individual. Recent developments in oxidant biology indicate that antioxidant requirements need to be re-evaluated in relation to cellular dysfunction. The requirement for oxidant defense will vary with oxidant stress. Antioxidant effects defined under conditions of zero stress are unlikely to be meaningful for any system either under brief, acute stress such as viral infections, inflammation, trauma, and exposure to environmental pollutants or during states of chronic and sustained oxidant stress such as autoimmunity, chronic infection, elevated circulating lipoproteins, or mild antioxidant deficiency. Thus, estimated requirements for antioxidant defense of a population would seem to be best based on individuals exposed to an "average" or "typical" amount of oxidant stress. However, the concept of what constitutes typical oxidant stress is undefined, and such definition would be difficult because a variety of insults can elicit an oxidant stress both directly and indirectly in ways heretofore unrecognized.

DISEASES WHOSE CAUSES AND SEVERITY ARE LINKED WITH OXIDATION

Oxygen-free radicals have been implicated as mediators of degenerative and chronic deteriorative, inflammatory, and autoimmune diseases (Miesel and Zuber, 1993) such as rheumatoid arthritis (Heliovaara et al., 1994), diabetes, vascular disease and hypertension (Deucher, 1992; Harris, 1992), cancer and hyperplastic diseases (Ames et al., 1993; Ferguson, 1994), cataract formation (Ames et al., 1993; Gershoff, 1993), emphysema (Rice-Evans and Diplock, 1993), immune system decline, and brain dysfunction as well as the aging process (Ames et al., 1993). Radical-mediated pathologies such as ischemia reperfusion and asthma (Bast et al., 1991) involve an imbalance in oxidant-antioxidant activity.

As discussed above, oxidants or oxidation and perturbations of oxidant balance control a number of cellular and organismal processes. Multiple mechanisms leading to systemic, multifactorial diseases probably operate simultaneously, an excellent example being found in the "French Paradox."

FRENCH PARADOX—DIETARY FAT VERSUS DIETARY PHYTOCHEMICALS

For certain French populations, the association of saturated fatty acid intake with mortality and morbidity from coronary heart disease (CHD) does not apply. The reported coronary mortality per 10,000 people in the United States is 182, the overall mortality of the French is 102, and, in the Toulouse region, it is 78 (Renaud and de Lorgeril, 1992). This discrepancy is referred to as the French Paradox. The French population cited consumes similar amounts of saturated fats and has similar risk factors and comparable plasma cholesterol as the population in the United States. The one dietary factor that showed a negative correlation with CHD was consumption of wine. While there is an association between reductions in CHD and alcohol consumption generally, the "alcohol" in red wine was superior to that provided in other alcoholic beverages (Renaud and de Lorgeril, 1992). Thus, it appeared that components other than alcohol were at least partly responsible for the beneficial effects of red wine.

CHD AND OXIDATION OF LDL

To understand how phytochemicals could reduce the incidence of CHD, the pathogenesis of this disease must be understood. The lesions of CHD are progressive, that is, they become larger over time. The first identifiable lesion is the fatty streak, which is a cluster of lipid-engorged macrophages called foam cells. The lipids in these cells are largely derived from LDL, but how the macrophage, without an LDL receptor, accumulates such high levels of LDL was reconciled only recently by the oxidation hypothesis. Oxidative modification of LDL makes native, and possibly quite benign, LDL a substrate for the macrophage scavenger receptor and hence more atherogenic (Steinberg et al., 1989; Steinberg, 1992). Oxidized cholesteryl esters and other PUFA oxidation products modify LDL and contribute to the etiology of CHD (Steinberg et al., 1989). The increased atherogenicity of LDLs following oxidative modification is proposed to result from their uncontrolled uptake into subendothelial macrophages by a receptor specific for oxidized particulates, such as red blood cells, and LDL (Sambrano et al., 1994). The accumulation of oxidized LDL by these macrophages gradually develops an inflamed and proliferative tissue site that leads ultimately to atheromatous plaque and vascular disease. A heart attack occurs when the roughened surface of the vascular tissue that overlies the plaque activates platelets, which form a clot or thrombus and occlude the artery. Platelet activation is also an oxidant-dependent reaction. Prevention of CHD by antioxidant protection could result from both prevention of peroxidative lipoprotein modification and the additional effects of antioxidants on cellular or immunological activity (Gey, 1990). There are several sites at which interference with oxidant generation and its consequences would slow CHD or reduce the incidence of heart attacks. Antioxidants, including some phytochemicals, may reduce peroxidation of PUFA and LDL and thereby decrease macrophage foam cell formation. They could reduce chronic inflammation tendencies by reducing peroxides, down-regulating the arachidonic acid cascade, and decreasing thrombotic tendencies and the ability of platelets to aggregate (Kinsella et al., 1993).

The view that oxidation as the key to CHD is somewhat at odds with traditional views of atherosclerotic risk. Elevated LDL is the fundamental correlate of CHD incidence and has captured considerable attention due to its ability to predict death from CHD in humans. Only recently was an explanation proposed for why, if LDL oxidation was causing CHD, high levels of LDL should be more readily oxidized. Furthermore, why should saturated fats that are less susceptible to oxidation promote atherosclerosis? A mechanistic linkage between increased LDL and LDL oxidation was provided by the recognition that increased saturated fats in the diet leads to a down-regulation of the LDL receptor in the liver. Since the liver is the major site of removal of LDL from blood, this has the net effect of increasing the amount of time that each LDL particle spends in the circulation. The susceptibility to oxidation of lipoproteins in circulation was shown to increase as their age in circulation increases (Walzem et al., 1995). Because increased intravascular LDL residence time in humans almost invariably accompanies elevated plasma LDL cholesterol (Stacpoole et al., 1991; Eriksson et al., 1993), the implication is that, on the average, the LDL from plasma of individuals with elevated plasma cholesterol is older than LDL isolated from plasma of individuals with normal plasma cholesterol. Thus, individuals with high LDL cholesterol impose an elevated oxidant stress on LDL by prolonging exposure to the oxidative environment of the intravascular and subendothelial compartments (Walzem et al., 1995).

If protection of LDL from oxidant stress prevents or slows the development of CHD, it becomes important to determine how and what can protect LDL. Vitamin E has a well-recognized physiological function as the major lipid-soluble biological antioxidant that scavenges free radicals (Burton and Ingold, 1989) and prevents oxidant injury to PUFA in cell membranes or within the surface of lipoproteins. LDL is the metabolic end-product of very-low-density lipoprotein (VLDL) catabolism. The number of tocopherol molecules present in each LDL particle is determined by the activity of the hepatic tocopherol transfer protein during synthesis of VLDL. Once secreted, if the protectors of lipoprotein are consumed, they are not likely to be replenished. An LDL particle contains fewer than ten tocopherol molecules (Frei and Gaziano, 1993). Whether this amount of tocopherol as the sole antioxidant protection available to the LDL particle is adequate or reflective of the conditions that humans evolved with is not known.

PHYTOCHEMICALS AS ANTIOXIDANTS

The relative role of different antioxidants in protection of LDL is still controversial. Some of the controversy stems from differences in methodologies that have been used. Water-soluble antioxidants, such as acorbate, in plasma appear to be the most effective protectors (Frei et al., 1988). However, the preparation of LDL via centrifugation and dialysis removes water-soluble antioxidants, therefore LDLs *in vitro* are protected from oxidation solely by lipid-soluble antioxidants such as vitamin E. Although tocopherols effectively delimit propagation of free radical chain reactions and may be partially regenerated by other reductants, their actions as sole protectants may be limited as they do not stop free radical initiation reactions.

Protection of the French from CHD has been hypothesized (Frankel et al., 1993; Kanner et al., 1987; Steinberg, 1992) to be a result of consumption of phenolic antioxidants that protect against LDL oxidation and inhibit platelet aggregation. Consumption by the French of a diet containing phytochemical antioxidants may decrease the perioxidative tendencies and retard interactions involved in atherogenesis and thrombosis. Phytochemicals can effectively participate in several antioxidant defenses, inhibit platelet aggregation, and spare α-tocopherol; they may protect sensitive targets such as proteins or DNA.

There is a lack of knowledge about the molecular composition of naturally occurring antioxidants, the amount of active ingredients in the source material, and relevant toxicity data. It was proposed that altered disease risk in specific populations could be explained by a mechanism involving antioxidant polyphenolics in fruits and fruit products, and this possibility focused considerable attention on the actions of these compounds in human health. This interest has produced epidemiological studies, hypotheses for mechanisms of action, and testing of oxidant/antioxidant effects in the progression of several diseases that can be classified by the aberration in oxidant balance that is believed to cause them. The breadth of associations between consumption of phenolic phytochemicals and improved human health emphasizes the need for further scientific investigation. The research group of German has focused on candidate molecules and their absorption and mechanisms of action (Frankel et al., 1993; Kanner et al., 1987; Kanner et al., 1994; Walzem et al., 1995). The research community is now developing testable hypotheses to further assess the mechanisms of these associations.

OXIDATION CAN DIRECTLY DAMAGE CERTAIN TARGETS AND PROMOTE DISEASE

DNA DAMAGE AND CANCER

Cancer is the second most significant chronic disease associated with direct oxidative damage. A review (Steinmetz and Potter, 1991a) of epidemiological literature reported a consistent association between higher levels of fruit and vegetable consumption and a reduced risk of cancer, particularly with epithelial cancers of the alimentary and respiratory tracts. Another review (Steinmetz and Potter, 1991b) addressed possible mechanisms by which these foods might alter risk of cancer. It was hypothesized that humans have adapted to a high intake of plant foods that supply crucial substances. Cancer may result from decreasing the level of intake of foods that are metabolically necessary. Among the potentially anticarcinogenic agents in foods are carotenoids, vitamin C, vitamin E, selenium, dietary fiber, dithiolthiones, glucosinolates, indoles, isothiocyanates, flavonoids, phenols, protease inhibitors, sterols, allium compounds, and limonene. The complementary and overlapping mechanisms suggested for these compounds (Steinmetz and Potter, 1991b) include "the induction of detoxification enzymes, inhibition of nitrosamine formation, provision of substrate for formation of antineoplastic agents, dilution and binding of carcinogens in the digestive tract, alteration of hormone metabolism, antioxidant effects, and others."

The biochemical actions of cancer-protective factors in fruits and vegetables were reviewed in detail (Dragsted et al., 1993). The review presented a simplified model on a generalized initiation-promotion-conversion model for carcinogenesis in which initiators are directly or indirectly genotoxic, promoters are substances capable of inferring a growth advantage on initiated cells, and converters are genotoxic. The mechanisms of anticarcinogenic substances in fruits and vegetables were related to the prevention and inhibition of cancer, notably by antioxidant-related activities. Polyphenols from fruits and vegetables were noted to protect against cancer initiation by scavenging activated mutagens and carcinogens, acting as antioxidants, inhibiting activating enzymes, and shielding sensitive molecules (e.g., DNA). Mechanisms acting at the biochemical level toward antipromotion include the scavenging of activated oxygen, stabilization of membranes, and inhibition of ornithine decarboxylase.

OXIDANT SIGNALING CAN PROMOTE CERTAIN DISEASES

Recognition that cellular stimulation activates oxidant production systems to generate very potent signal molecules has suggested the possibility that such systems could be inappropriately stimulated (German and Hu, 1990; German and Kinsella, 1986; Kanner et al., 1987). Thus, the systems that respond to stress may be destructive and exacerbate or even initiate distinct pathological states. The oxygenated fatty acids or eicosanoids are the best described of the known stress response systems, and several examples of pathologies associated with their excessive production are described below.

PROSTAGLANDINS AND THROMBOSIS

Inappropriate blood clotting is the accumulation of many factors that produce an occluded artery. The critical principle underlying this pathology of thrombosis, and the practical problem in successful interventions, is that platelet clotting is the essential event that prevents excessive bleeding. The vascular system is a network that carries a relatively viscous fluid at high flow and pressure, and the utility of platelet aggregation is obvious. Platelets act directly to recognize vessel wall disruptions, signal a rapid, multicellular response, and then initiate construction of a physical barrier to blood loss.

In view of the above-discussed paradigm, it is not surprising that oxidant signaling systems have evolved to respond rapidly to stress. Platelets are known to oxygenate arachidonic acid via two different oxygenating systems: the prostaglandin synthetase-thromboxane synthetase couple that produces thromboxane, the most active and potent platelet aggregating and vasoconstricting of the arachidonic acid metabolites, and the 12-lipoxygenase that produces 12(S)-HPETE, an arachidonic acid hydroperoxide that promotes adherence of platelets to vascular surfaces. Although these enzymes are not absolutely required elements of the platelet-clotting cascade, these two systems of oxygenating fatty acids accelerate and amplify the clotting processes. In response to both appropriate (bleeding) and inappropriate (thrombosis) conditions, platelets produce oxidants as a means to cascade the signaling of clot-

ting. Based on this knowledge, pharmacological agents that inhibit oxyge-
nases have been developed to prevent thrombosis, and many prostaglandin
inhibitors, including aspirin, are used as antithrombotic agents. Aspirin is a
salicylate phenol originally derived from willow (*Salix*) bark. The therapeutic
successes of small daily doses of aspirin suggest the possibility that polyphenolic phytochemicals consumed in foods may inhibit the enzymatic oxygenation of platelets that promote thrombosis. Epithelial lipoxygenase is inhibited
in vitro by micromolar amounts of catechin (Hsieh et al., 1988). Human
platelet lipoxygenase was inhibited by wine phenolics and catechin in pure
form (Matsuo and German, unpublished).

LEUKOTRIENES AND ASTHMA

The excessive recruitment of inflammatory immune cells in asthma is distinctive for this autoimmune disease. Several immune cell modulators are responsible for the bronchoconstriction associated with allergen challenge to
pulmonary mast cells. Leukotriene-5-lipoxygenase products are the most potent bronchoconstricting substances. Analogous to other signaling systems,
challenged cells release arachidonic acid, which is actively oxygenated via a
free radical oxidation reaction to a stereospecific hydroperoxide. Hydroperoxide from leukotrienes is converted by an additional free radical reaction to the
leukotriene precursor, LTA4. In asthma, mast cells are considered the main
source of histamine and slow-reacting substances of anaphylaxis, SRS-A.
SRS-A is the signal molecule implicated in bronchoconstriction, but neutrophils and eosinophils are also associated with asthmatic lungs. The recruitment and activation of these cells is enhanced by oxidized lipids, especially
the eicosanoids, and cell activity and severity of bronchoconstriction are decreased by pharmacological agents that block oxidized signal lipids. Various
phytochemicals have been documented to inhibit these pathways, and they
have been argued to be of therapeutic benefit for inflammatory lung diseases.

CELL ACTIVATION BY OXIDANTS CAN PROMOTE CERTAIN DISEASES

PHAGOCYTE ACTIVATION AND ASBESTOSIS

Observations suggest that neutrophils may play a role in the development
of lung injury after exposure to asbestos. Macrophages attracted by lung mediators phagocytize asbestos fibers. Asbestos also exerts a cytotoxic effect on
polymorphonuclear leukocytes, which then release proteolytic enzymes such
as collgenase and elastase. Highly reactive oxidants produced in neutrophils
in response to asbestos cause additional damage to tissues. While the role of
polymorphonuclear leukocytes, such as neutrophils, in the pathogenesis of

asbestos remains controversial (Lewczuk et al., 1994), the potential for antioxidant phytochemicals as protectants from direct oxidant damage and for down-regulating oxidant signaling in recruiting phagocytic cells is evident. The beneficial effects of tocopherol against asbestos-induced oxidant damage pose the questions "Is this a paradigm of phenolic protection from excessive inflammation?" and "Could other phytochemicals modify aggressive activation and oxidant damage during phagocytosis?"

UNRESOLVED ISSUES

While plant phytochemicals clearly hold promise for preventing and ameliorating disease, important scientific questions are "Which compounds are effective" and "What is their mechanism of action?" Some key issues that must be addressed to answer the questions are (1) identification and quantitation of phytochemicals in foods, (2) determination of the bioavailability to humans and pharmacokinetics of specific compounds, (3) determination of the molecular structure of phytochemicals in relation to antioxidant actions, redox properties, enzyme alterations, and metabolic responses, (4) identification and quantitation of phytochemicals and their metabolites in clinical samples from humans, (5) determination of molecular targets of action and doses required to provide efficacy, and (6) determination of the molecular mechanisms of disease intervention.

As this review has illustrated, the chemistry of biological response to oxidation are complex. Partially as a result of this complexity, much of the published data addressing the above-stated issues is questionable. The approach most likely to expand the understanding of oxidation and health is to measure specific compounds in tissues and biological fluids. Past problems in achieving this goal are related to the fact that quantitation of phenolics in various foods was based upon measurements of total classes of polyphenols as opposed to specific polyphenols. Inaccuracies have resulted from use of unsophisticated methodology for identification and quantitation. Much of this older methodology has been replaced by analytical techniques with the lower detection levels required for quantitation of minor, but potentially biologically active, compounds. An additional concern relates to the most appropriate methods to handle clinical samples to be measured for the appearance or disappearance of polyphenols absorbed following ingestion or other means of administration. Studies have yet to determine the most appropriate methods for collection, transport, and storage of the large variety of clinical samples (i.e., plasma versus serum, temperature, and binding of polyphenols to proteins). Future studies need to determine the effects of the food matrix of polyphenol-containing materials on the absorption of these compounds.

Epidemiology has provided support for a positive outcome of future research on phytochemicals and health. Population correlations must now be transformed into research on specific molecules, mechanisms of action, biomarkers of their status and functions, efficacious delivery, and means to improve the food supply to realize the value of this scientific scrutiny.

ACKNOWLEDGEMENTS

The authors acknowledge the support of the United States – Israel Binational Agricultural Research and Development Fund.

REFERENCES

Ames, B. N., Shigenaga, M. K., and Hagen, T. M. 1993. Oxidants, Antioxidants, and the Degenerative Diseases of Aging. *Proc. Natl. Acad. Sci. USA,* 90:7915 – 7922.

Bashir, S., Harris, G., Denman, M. A., Blake, D. R., and Winyard, P. G. 1993. Oxidative DNA Damage and Cellular Sensitivity to Oxidative Stress in Human Autoimmune Disease. *Ann. Rheum. Dis.,* 52:659 – 666.

Bast, A., Haenen, G. R., and Doelman, C. J. 1991. Oxidants and Antioxidants: State of the Art. *Am. J. Med.,* 91(3C):2S – 13S.

Buettner, G. R. 1993. The Pecking Order of Free Radicals and Antioxidants: Lipid Peroxidation, alpha-Tocopherol, and Ascorbate. *Arch. Biochem. Biophys.,* 300:535 – 543.

Burton, G. W., and Ingold, K. 1989. Vitamin E as an *In Vitro* and *In Vivo* Antioxidant. *Ann. New York Acad. Sci.,* 570:7 – 22.

Deucher, G. P. 1992. Antioxidant Therapy in the Aging Process. *Exs,* 62:428 – 437.

Dragsted, L. O., Strube, M., and Larsen, J. C. 1993. Cancer-Protective Factors in Fruits and Vegetables: Biochemical and Biological Background. *Pharmacol. Toxicol.* 72 (Suppl 1):116 – 135.

Eriksson, M., Berglund, L., Gabrielsson, J., Lantz, B., and Angelin, B. 1993. Non-steady-state Kinetics of Low Density Lipoproteins in Man: Studies after Plasma Exchange in Healthy Subjects and Patients with Familial Hypercholesterolaemia. *Eur. J. Clin. Invest.,* 23:746 – 752.

Ferguson, L. R. 1994. Antimutagens as Cancer Chemopreventive Agents in the Diet. *Mutat. Res.,* 307:395 – 410.

Frankel, E., Kanner, J., German, J. B., Parks, E., and Kinsella, J. E. 1993. Inhibition of Oxidation of Human Low-Density Lipoprotein by Phenolic Substances in Red Wine. *Lancet,* 341:454 – 457.

Frankel, E. N. 1991. Recent Advances in Lipid Oxidation. *J. Sci. Food Ag.,* 54:495 – 511.

Frei, B., and Gaziano, J. M. 1993. Content of Antioxidants, Preformed Lipid Hydroperoxides, and Cholesterol as Predictors of the Susceptibility of Human LDL to Metal Ion-dependent and Independent Oxidation. *J. Lipid Res.,* 34:2135 – 2145.

Frei, B., Stocker, R., and Ames, B. N. 1988. Antioxidant Defenses and Lipid Peroxidation in Human Blood Plasma. *Proc. Natl. Acad. Sci. USA,* 85:9748 – 9752.

Fritsche, K. L., and Johnston, P. V. 1988. Rapid Autoxidation of Fish Oil in Diets without Added Antioxidants. *J. Nutr.,* 118:425–426.

German, J. B., and Hu, M.-L. 1990. Oxidant Stress Inhibits the Endogenous Production of Lipoxygenase Metabolites in Rat Lungs and Fish Gills. *Free Radical Biol. Med.,* 8:441–448.

German, J. B., and Kinsella, J. E. 1986. Hydroperoxide Metabolism in Trout Gill Tissue: Effect of Glutathione on Lipoxygenase Products Generated from Arachidonic Acid and Docosahexaenoic Acid. *Biochim. Biophys. Acta,* 879:378–387.

Gershoff, S. N. 1993. Vitamin C (Ascorbic Acid): New Roles, New Requirements? *Nutr. Rev.,* 51:313–326.

Gey, K. F. 1990. The Antioxidant Hypothesis of Cardiovascular Disease: Epidemiology and Mechanisms. *Biochem. Soc. Trans.,* 18:1041–1045.

Gutteridge, J. M. 1994. Biological Origin of Free Radicals, and Mechanisms of Antioxidant Protection. *Chem.-Biol. Interact.,* 91:133–140.

Halliwell, B. 1991. Reactive Oxygen Species in Living Systems: Source, Biochemistry, and Role in Human Disease. *Am. J. Med.,* 91(3C):14S–22S.

Halliwell, B., and Gutteridge, J. M. C. 1989. *Free Radicals in Biology and Medicine.* Oxford: Clarendon Press.

Hanasaki, Y., Ogawa, S., and Fukui, S. 1994. The Correlation between Active Oxygen Scavenging and Antioxidative Effects of Flavonoids. *Free Radical Biol. Med.,* 16:845–850.

Harris, W. S. 1992. The Prevention of Atherosclerosis with Antioxidants. *Clin. Cardiol.,* 15:636–640.

Heliovaara, M., Knekt, P., Aho, K., Aaran, R. K., Alfthan, G., and Aromaa, A. 1994. Serum Antioxidants and Risk of Rheumatoid Arthritis. *Ann. Rheum. Dis.,* 5:51–53.

Hsieh, R. J., German, J. B., and Kinsella, J. E. 1988. Relative Inhibitory Potencies of Flavonoids on 12-Lipoxygenase of Fish Gill. *Lipids,* 23:322–326.

Huang, S.-W., Frankel, E. N., and German, J. B. 1994. Antioxidant Activity of Alpha- and Gamma-Tocopherols in Bulk Oils and in Oil-in-Water Emulsions. *J. Agric. Food Chem.,* 42:2108–2114.

Huang, S.-W, Frankel, E. N., and German, J. B. 1995. Effects of Individual Tocopherols and Tocopherol Mixtures on the Oxidative Stability of Corn Oil Triglycerides. *J. Agric. Food Chem.,* 43:2345–2350.

Jessup, W., Rankin, S. M., De Whalley, C. V., Hoult, J. R., Scott, J., and Leake, D. S. 1990. Alpha-Tocopherol Consumption During Low-Density-Lipoprotein Oxidation. *Biochem. J.,* 265:399–405.

Kanner, J., Frankel, E., Granit, R., German, B., and Kinsella, J. E. 1994. Natural Antioxidants in Grapes and Wines. *J. Ag. Food Chem.,* 42:64–69.

Kanner, J., German, J. B., and Kinsella, J. E. 1987. Initiation of Lipid Peroxidation in Biological Systems. *Crit. Rev. Food Sci. Nutr.,* 25:317–364.

Keen, C. L., German, J. B., Mareschi, J. P., and Gershwin, M. E. 1991. Nutritional Modulation of Murine Models of Autoimmunity. *Rheum. Dis. Clin. North Am.,* 17:223–234.

Kinsella, J. E., Frankel, E., German, B., and Kanner, J. 1993. Possible Mechanisms for the Protective Role of Antioxidants in Wine and Plant Foods. *Food Technol.,* 47:85–90.

Laughton, M. J., Evans, P. J., Moroney, M. A., Hoult, J. R., and Halliwell, B. 1991. Inhibition of Mammalian 5-Lipoxygenase and Cyclo-oxygenase by Flavonoids and Phenolic Dietary Additives. Relationship to Antioxidant Activity and to Iron Ion-reducing Ability. *Biochem. Pharmacol.,* 42:1673–1681.

Levine, R. L., Mosoni, L., Berlett, B. S., and Stadtman, E. R. 1996. Methionine Residues as Endogenous Antioxidants in Proteins. *Proc. Natl. Acad. Sci. USA,* 93:15036–15040.

Lewczuk, E., Owczarek, H., and Staniszewska, G. 1994. Contribution and Role of Polymorphonuclear Leucocytes in Inflammatory Reactions of Asbestosis. *Medycyna Pracy,* 45:547–550.

Miesel, R., and Zuber, M. 1993. Copper-Dependent Antioxidase Defenses in Inflammatory and Autoimmune Rheumatic Diseases. *Inflammation,* 17:283–294.

Muggli, R. 1989. Dietary Fish Oils Increase the Requirement for Vitamin E in Humans. In *Health Effects of Fish and Fish Oils.* Chandra, R. K., ed. St. John's, Newfoundland, Canada: ARTS Biomedical Publishers and Distributors Ltd., pp. 201–210.

Porter, N. A., Caldwell, S. E., and Mills, S. A. 1995. Mechanisms of Free Radical Oxidation of Unsaturated Lipids. *Lipids,* 30:277–290.

Renaud, S., and de Lorgeril, M. 1992. Wine, Alcohol, Platelets, and the French Paradox for Coronary Heart Disease. *Lancet,* 339:1523–1526.

Rice-Evans, C. A., and Diplock, A. T. 1993. Current Status of Antioxidant Therapy. *Free Radical Biol. Med.,* 15:77–96.

Sambrano, G. R., Parthasarathy, S., and Steinberg, D. 1994. Recognition of Oxidatively Damaged Erythrocytes by a Macrophage Receptor with Specificity for Oxidized Low Density Lipoprotein. *Proc. Natl. Acad. Sci. USA,* 91:3265–3269.

Stacpoole, P. W., von Bergmann, K., Kilgore, L. L., Zech, L. A., and Fisher, W. R. 1991. Nutritional Regulation of Cholesterol Synthesis and Apolipoprotein B Kinetics: Studies in Patients with Familial Hypercholesterolemia and Normal Subjects Treated with a High Carbohydrate, Low Fat Diet. *J. Lipid Res.,* 23:746–1848.

Steinberg, D. 1992. Antioxidants in the Prevention of Human Atherosclerosis. *Circulation,* 85:2337–2344.

Steinberg, D., Parthasarathy, S., Carew, T. E., Khoo, J. C., and Witztum, J. L. 1989. Beyond Cholesterol. Modifications of Low-Density Lipoprotein That Increase Its Atherogenicity. *New Engl. J. Med.,* 320:915–924.

Steinmetz, K. A., and Potter, J. D. 1991a. Vegetables, Fruit, and Cancer. I. Epidemiology. *Cancer Causes Control,* 2:325–357.

Steinmetz, K. A., and Potter, J. D. 1991b. Vegetables, Fruit, and Cancer. II. Mechanisms. *Cancer Causes Control,* 2:427–442.

Stohs, S. J., and Bagchi, D. 1995. Oxidative Mechanisms in the Toxicity of Metal Ions. *Free Radical Biol. Med.,* 18:321–336.

Suzuki, Y. J., Mizuno, M., Tritschler, H. J., and Packer, L. 1995. Redox Regulation of NF-kappa B DNA Binding Activity by Dihydrolipoate. *Biochem. Mol. Biol. Internat.,* 36:241–246.

Terao, J., Piskula, M., and Yao, Q. 1994. Protective Effect of Epicatechin, Epicatechin Gallate, and Quercetin on Lipid Peroxidation in Phospholipid Bilayers. *Arch. Biochem. Biophys.,* 308:278–284.

Tournaire, C., Croux, S., Maurette, M. T., Beck, I., Hocquaux, M., Braun, A. M., and Oliveros, E. J. 1993. Antioxidant Activity of Flavonoids: Efficiency of Singlet Oxygen (1 delta g) Quenching. *Photochem. Photobiol. B Biol.,* 19:205–215.

Walzem, R. L., Watkins, S., Frankel, E. N., Hansen, R. J., and German, J. B. 1995. Older Plasma Lipoproteins are More Susceptible to Oxidation: a Linking Mechanism for the Lipid and Oxidation Theories of Atherosclerotic Cardiovascular Disease. *Proc. Natl. Acad. Sci. USA,* 92:7460–7464.

CHAPTER 3

Tocotrienols in Metabolism

ROSS L. HOOD

INTRODUCTION

Iɴ the development of coronary heart disease, hypercholesterolemia, which is an elevation in low-density lipoprotein (LDL) cholesterol concentration, is a major risk factor. The American Heart Association's Task Force on Cholesterol Issues (LaRosa et al., 1990) suggested that "the evidence linking serum cholesterol to coronary heart disease is overwhelming." A graded relationship exists between serum cholesterol concentration and the incidence of coronary heart disease (Palca, 1990). This author suggests that, in general, a 1% fall in serum cholesterol is associated with at least a 2% fall in the incidence of the disease. He also concludes that there will be a considerable reduction in heart disease if plasma cholesterol is lowered in the whole community rather than just those adults with a particularly high plasma cholesterol.

The relationship between cholesterol metabolism and dietary tocotrienols was reported by Qureshi et al., in 1986. This group of researchers was the first to demonstrate that tocotrienols inhibited the activity of 3-hydroxy-3-methylglutaryl coenzyme A (HMGCoA) reductase, the rate-limiting enzyme in cholesterol synthesis. Their initial discovery has triggered a number of studies to clarify the role of these biologically active compounds in cholesterol metabolism. Natural palm oil and its commercial fractions (e.g., palm olein) are excellent sources of vitamin E (α-tocopherol) and compounds with vitamin E activity (the tocotrienols). Palm oil is the only commercial

vegetable oil to contain significant amounts of tocotrienols; the tocol content
of palm oil is typically between 500 and 750 ppm (Goh et al., 1985).

BIOCHEMISTRY

The term *vitamin E* is a collective name for naturally occurring tocopherols
and tocotrienols that are present in many plants, particularly those plants with a
high oil content. The structure of both tocopherols and tocotrienols is characterized by a chromanol head group and a phytyl side chain. The tocopherol can be
designated α, β, γ, or δ, depending on the position of the methyl substitutions on
the aromatic ring of the chromanol head group. The only difference in the struc-
ture of tocotrienols and tocopherols is in the phytyl tail; tocotrienols have three
unsaturated double bonds in the tail, whereas tocopherols have a saturated tail
(Figure 1).

The phytyl tail of tocopherols and tocotrienols binds these compounds into
lipoproteins and membranes, where they coexist with polyunsaturated fatty
acids. Oxidation of the latter will result in lipid peroxidation to form hydroperoxides and a series of lipid radicals. Once peroxidation is initiated, a
chain of free radical reactions is triggered, which can destroy lipids and nu-
cleic acids that may be in close proximity to free radicals. Vitamin E can
break this chain reaction at the propagation stage of free radical reactions
(Burton and Ingold, 1989), and this is an important step in the propagation of
many chronic and degenerative diseases of aging. Understanding the differ-

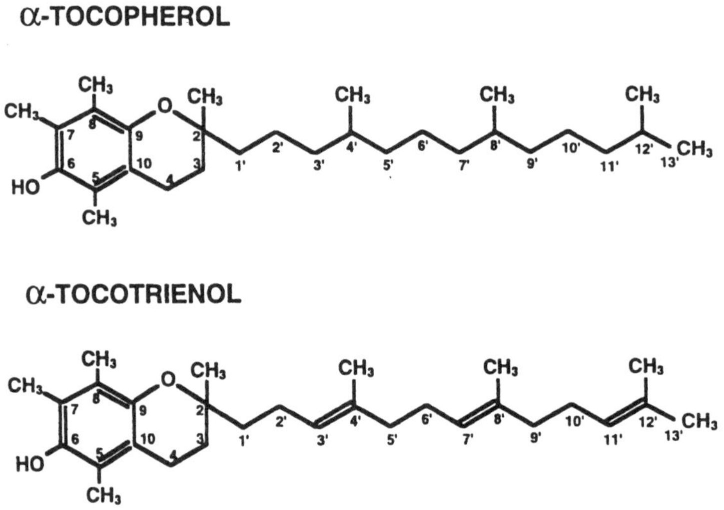

α-TOCOPHEROL

α-TOCOTRIENOL

Figure 1 Molecular structure of α-tocopherol and α-tocotrienol (Packer, 1995).

ence in the structure of the phytyl chain is important in explaining the difference in antioxidant activities of tocopherols and tocotrienols.

The antioxidant activities of α-tocopherol and α-tocotrienol in isolated microsomal suspensions using different systems (e.g., ascorbate + Fe^{2+} or NADPH + Fe^{2+}) for initiating lipid peroxidation have been studied. Serbinova et al. (1992) observed that α-tocotrienol was more effective than α-tocopherol in the inhibition of lipid peroxidation in these systems. The concentrations of α-tocopherol producing 50% inhibition (K_{50}) were 40 and 60 times higher that those for α-tocotrienol for (Fe^{2+} +NADPH) and (Fe^{2+} +ascorbate)-dependent lipid peroxidation, respectively (Serbinova et al., 1992). Packer (1995) compared the antioxidant potencies of these two tocols in a more physiological system [Fe(II) + ascorbate or Fe(II) + NADPH induced lipid peroxidation in rat liver microsomes]. He reported that α-tocotrienol exerted greater antioxidant activity than α-tocopherol. Packer's, 1995 review provides an excellent profile of antioxidant properties of tocols.

ANIMAL STUDIES

When pigs, either hypercholesterolemic or normolipemic, were fed a balanced diet that was supplemented with 50 $\mu g/g$ of a tocotrienol-rich fraction (TRF) isolated from palm oil, the hypercholesterolemic pigs showed a 44% reduction in total serum cholesterol and a 60% reduction in low density lipoprotein (LDL) cholesterol (Qureshi et al., 1991b). These authors suggested that TRF has a protective effect on the endothelium and on platelet aggregation since they observed a 41% decline in thromboxane B_2 and a 29% reduction in platelet factor 4. Biopsied samples of adipose tissue were taken from all pigs, and a significant ($p < 0.01$) reduction in HMGCoA reductase activity was reported in both groups of pigs supplemented with tocotrienols. It is also useful to note that the hypocholesterolemic effect of TRF persisted only in the hypercholesterolemic swine for 8 weeks after they were placed back on their control diet.

TRF from palm oil at varying concentrations (0–1000 ppm) has been fed to male Wistar rats for a 3-week period (Hood, 1995). TRF was included in the diet by dissolving it in acetone and spraying it onto commercial rat cubes. Tocols were assayed in the blood, and the data (Table 1) indicated that dietary tocols were transferred to the blood plasma. However, when tocols were administered to rats in this way, they were ineffective in producing any consistent changes in blood cholesterol or in hepatic cholesterol synthesis. In a related study, Hood (1995) gave TRF to rats that were fed cholesterol and cholate (Table 2). Dietary cholesterol and cholate significantly increased blood and liver cholesterol and decreased *in vivo* cholesterol synthesis. The

TABLE 1. Effect of Dietary Cholesterol and Tocotrienols from Palm Oil on Cholesterol Metabolism of Male Wistar Rats.

	Dietary TRF (ppm)			
	0	50	250	1000
Live weight (g)	238.7 ± 3.7	250.1 ± 2.2	258.7 ± 9.3	253.5 ± 5.5
Liver weight (g)	9.4 ± 0.2	10.1 ± 0.6	10.4 ± 0.4	10.7 ± 0.3
Liver cholesterol (mg/liver)	18.6 ± 0.4	21.8 ± 1.2	21.8 ± 1.2	21.2 ± 0.8
Blood cholesterol (mmol/L)	1.97 ± 0.13	1.86 ± 0.11	1.75 ± 0.12	2.04 ± 0.08
HDL cholesterol (mmol/L)	0.94 ± 0.08	0.84 ± 0.06	0.86 ± 0.06	0.95 ± 0.04
Cholesterol synthesis[a]	331.9 ± 41.8	467.6 ± 43.0	319.0 ± 33.6	478.5 ± 51.9
Plasma α-tocopherol (μmol/L)[b]	1.0	9.0	21.5	14.0
Plasma α-tocotrienol (μmol/L)[b]	0.4	0.5	1.0	4.7
Plasma γ-tocotrienol (μmol/L)[b]	0.3	0.4	1.0	4.7

[a]Nanomoles of ^{14}C-mevalonate to cholesterol/hour/liver.
[b]Assayed by Roland Stocker, The Heart Research Institute, Sydney, Australia.
Source: Adapted from Hood (1995).

TABLE 2. Effect of Dietary Cholesterol and Tocotrienols from Palm Oil on Cholesterol Metabolism of Male Wistar Rats.

	A	B	C
Live weight (g)	154.5 ± 7.9	153.5 ± 7.3	147.3 ± 6.9
Liver weight (g)	7.2 ± 10.4	9.2 ± 0.5	8.0 ± 0.6
Liver cholesterol (mg/liver)	16.3 ± 1.0	373.1 ± 33.1	247.9 ± 14.7
Blood cholesterol (mmol/L)	3.11 ± 0.13	8.89 ± 0.79	8.29 ± 1.09
HDL cholesterol (mmol/L)	1.13 ± 0.04	0.52 ± 0.06	0.77 ± 0.05
LDL cholesterol (mmol/L)	1.98 ± 0.11	8.37 ± 0.74	7.52 ± 1.00
Cholesterol synthesis[b]	1029 ± 58	233 ± 32	197 ± 38
Cholesterol synthesis ($\times 10^{-2}$)[c]	399 ± 86	89 ± 6	71 ± 13

[a]The diets comprise crushed rat cubes, plus 8% palm olein plus
  A = plus 250 ppm TRF or
  B = plus 1% cholesterol + 0.2% cholate or
  C = plus 250 ppm TRF + 1% cholesterol + 0.2% cholate
[b]Nanomoles of ^{14}C-mevalonate to cholesterol/hour/liver.
[c]dpm ^{3}H$_2$O to cholesterol/hour/liver.
Source: Adapted from Hood (1995).

addition of tocols (250 ppm) did not alter blood cholesterol; however, the to-cols caused a reduction in liver cholesterol synthesis, which was reflected in a significantly lower liver cholesterol concentration. These two studies indicate that tocotrienols do not bring about a consistent cholesterolemic response in rats. In contrast, another study (Qureshi et al., 1989) reported that a TRF from barley oil suppressed HMGCoA reductase activity in hepatocytes in a dose-dependent manner.

HUMAN STUDIES

Palmvitee capsules, a product developed by the Palm Oil Research Institute of Malaysia, are typically the source of dietary tocols used in human studies. These capsules contain approximately 50 mg of TRF from palm oil dissolved in 250 mg of palm olein. In one study (Tan et al., 1991), 31 subjects were each given one Palmvitee capsule for 30 days. Their serum cholesterol decreased from 6.48 to 5.43 mmol/L. The greatest reduction occurred in the LDL cholesterol fraction, and individuals with a high initial blood cholesterol were the most responsive to dietary tocols. In an American study (Qureshi et al., 1991c) where four capsules of Palmvitee were given daily to 25 hyper-cholesterolemic subjects, a nonsignificant hypocholesterolemic effect was observed. These same researchers compared the effect of Palmvitee and γ-tocotrienol on blood cholesterol in 40 hypercholesterolemic patients under controlled dietary conditions. Subjects given Palmvitee and γ-tocotrienol capsules had their blood cholesterol reduced by 15 and 20%, respectively (Qureshi et al., 1995). They noted individual differences and that 20% of the subjects did not respond to the dietary tocols. This team of research workers indicate that LDL-cholesterol concentrations are linked to the quantity of tocols bound to the LDL moiety and not the concentration of free tocols in the serum (Qureshi et al., 1996a).

An Australian study (Wahlqvist et al., 1992) with 44 subjects investigated the effect of a tocotrienol-rich vitamin E from palm oil on serum tocols, serum lipids, platelet function, and prostaglandins. Tocol concentrations increased and plateaued in the serum by the fourth week; however, at no stage was there any significant change in serum LDL cholesterol, HDL cholesterol, or triacylglycerols.

TOCOTRIENOLS AND ATHEROSCLEROSIS

In a study of the antioxidant properties of Palmvitee, Tomeo et al. (1995) measured, over an 18-month period, serum lipids, platelet aggregation, fatty acid peroxides, and carotid artery stenosis in 50 patients with cerebrovascular disease. Bilateral duplex ultrasonography revealed apparent carotid atherosclerotic regression in 7 and progression in 2 of the 25 tocotrienol patients, while none of the control group exhibited regression and 10 of 25 showed progression ($p < 0.002$). Serum total cholesterol, LDL cholesterol, and triacylglycerols remained unchanged in both groups, whereas serum thiobarbituric acid-reactive substance, an *ex vivo* indicator of maximal platelet peroxidation, decreased in the tocotrienol group ($p < 0.05$) and increased nonsignificantly in the control group. These findings led Tomeo et al. to conclude that antioxidants, such as tocotrienols, may influence the course of carotid atherosclerosis.

AVIAN STUDIES

Early avian studies were carried out with White Leghorn pullets fed either 5% refined, bleached, and deodorized palm olein or 5% corn oil (Elson, 1992). After feeding for 1 month, blood cholesterol concentrations were similar, whereas the activity of HMGCoA reductase was lower in the livers of the birds treated with palm olein. In a second study, White Leghorn layers were fed a diet containing 200 ppm of TRF. This tocol mixture consisted of 26% *d*-α-tocopherol, 2% *d*-g-tocopherol, 25% *d*-α-tocotrienol, 38% *d*-γ-tocotrienol, and 9% *d*-δ-tocotrienol. Blood cholesterol concentrations in layers fed the tocol mixture for 3 weeks were 12 to 27% lower than in birds fed an equivalent amount of α-tocopherol. This study confirms an earlier observation that supplemental α-tocopherol tends to elevate serum cholesterol (Qureshi et al.,

1989). The initial study that linked decreased hepatic cholesterolgenesis with dietary tocotrienols was carried out in chickens (Qureshi et al., 1986). In this study, the tocotrienols were fed at 2.5 to 20 ppm, and these workers were able to demonstrate that the cholesterol-suppressive action of tocotrienols is at the level of HMGCoA reductase.

When ^{3}H-γ-tocotrienol was fed to chickens and the serum tocols assayed for radioactivity, most of the radioactivity was found in α-tocopherol, with some radioactivity found in α-tocotrienol, γ-tocotrienol, and γ-tocopherol. No radioactivity was found in the δ-isomers, which supports the view that the same biosynthetic pathway exists in birds that has been proposed for plants. It is likely that avian gut microflora enzymes may be responsible for these conversions. Other studies have also observed an increase in blood α-tocopherol after feeding γ-tocotrienol.

Male Japanese quail are prone to atherosclerosis, which can be induced by the feeding of cholesterol (Donaldson, 1982; Shih, 1983). The cholesterol status of Japanese quail is also very responsive to dietary fats of different fatty acid composition (Hood, 1990, 1991). In these studies, when blood lipid concentrations and rates of hepatic cholesterol synthesis were compared in quail that were fed different fats, rates of hepatic synthesis mirrored the results for serum concentration of cholesterol.

In both animal (Evans et al., 1992) and human (Jenkins et al., 1975) studies, guar gum has been shown to be effective in lowering plasma cholesterol. The mechanism by which soluble dietary fiber has a hypocholesterolemic effect is quite different from that proposed for tocotrienols. Keeping these two mechanisms in mind, Hood and Sidhu (1992) looked for a synergistic effect of dietary fiber (guar gum or α-cellulose) and tocotrienols from palm oil on cholesterol status of young and mature Japanese quail. Dietary guar gum and tocotrienols (50 ppm) were effective in reducing plasma cholesterol and liver size of young quail when compared to birds fed α-cellulose. Tocotrienols were not effective in altering cholesterol metabolism in quail fed α-cellulose. Tocotrienols were also not effective in altering cholesterol metabolism in mature, 52-week-old quail.

A third experiment was carried out where palm oil tocotrienols were fed to male Japanese quail at either 0, 50, 250, or 1,000 ppm (Hood, 1995). The tocotrienol mixture was added to a crushed commercial turkey starter. Diet did not alter blood cholesterol (Table 3) whether measured by the Reflotron instrument or a standard enzymatic kit. Rates of cholesterol synthesis, as measured by the incorporation of either ^{3}H$_2$O or ^{14}C-mevalonate into cholesterol, were lower in birds from the groups fed the palm oil tocotrienols. In another study, male Japanese quails were orally administered with either 100 μL of canola oil or 100 μL of 5% TRF in canola oil. The basal diets were those described by Hood and Sidhu (1992) in an earlier experiment. When TRF was orally administered, total blood cholesterol decreased from 6.14 to 4.98

TABLE 3. Effect of Dietary Tocotrienol-Rich Fraction (TRF) Form Palm Oil on Cholesterol Metabolism of Male Japanese Quail.

	Dietary TRF (ppm)			
	0	50	250	1000
Live weight (g)	226.5 ± 8.1	238.3 ± 8.6	229.5 ± 6.2	230.5 ± 5.3
Liver weight (g)	3.17 ± 0.18	3.29 ± 0.24	3.27 ± 0.20	3.01 ± 0.19
Liver cholesterol (mg/liver)	9.1 ± 0.7	11.8 ± 2.4	10.2 ± 1.0	9.2 ± 0.5
Cholesterol synthesis[a]	58.2 ± 7.4	39.6 ± 5.7	49.3 ± 4.6	42.9 ± 5.9
Cholesterol synthesis ($\times 10^2$)[b]	322 ± 36	235 ± 22	270 ± 38	261 ± 35
Blood Cholesterol (mmol/L)[c]	5.33 ± 0.31	6.10 ± 0.67	5.14 ± 0.28	5.87 ± 0.22
HDL cholesterol (mmol/L)[c]	3.48 ± 0.31	4.27 ± 0.476	3.49 ± 0.24	4.05 ± 0.20
Blood cholesterol[d] (week 1)	5.5 ± 0.3	4.8 ± 0.6	5.3 ± 0.4	5.1 ± 0.3
(week 2)	5.8 ± 0.2	4.8 ± 0.6	5.3 ± 0.4	5.1 ± 0.3
(week 3)	5.0 ± 0.2	4.8 ± 0.5	4.5 ± 0.3	4.1 ± 0.3
(week 4)	5.2 ± 0.2	5.4 ± 0.5	5.1 ± 0.4	5.3 ± 0.2
(week 5)	5.0 ± 0.3	5.1 ± 0.6	5.1 ± 0.5	4.7 ± 0.4
(week 6)	4.6 ± 0.3	5.0 ± 0.6	4.8 ± 0.4	4.6 ± 0.3

[a]Nanomoles of ^{14}C-mevalonate to cholesterol/hour/liver.
[b]dpm ^3H_2O to cholesterol/hour/liver.
[c]Enzymatic kit.
[d]Reflotron results (mmol/L); TRF treatment commenced after week 2.
Source: Adapted from Hood (1995).

mmol/L, and liver cholesterol concentration and cholesterol synthesis rates from mevalonate were unchanged. The results to date with quails have been varied and inconclusive.

After discussions with an international researcher on tocotrienols, it was suggested that the inability to show an effect with previous studies on quails was due to the elevated fat content used in the diets. Hence, an experiment was designed to compare the effect of dietary fat on the cholesterolemic response to tocotrienols. Four levels of dietary fat were fed, the amount of corn oil was kept constant (0.55%), and equal portions of canola and coconut oil were varied (Table 4). Dietary fat, from vegetable origin, varied from 0.75 to 6.0% of the diet, and half the quails received 2 mg of TRF administered orally for a 3-week period. No significant effects on total or LDL blood cholesterol attributable to dietary TRF were observed (Table 5). Also, neither dietary fat nor dietary TRF affected liver cholesterol concentration. The activity of HMGCoA reductase was similar in all groups except when 1.5% dietary fat was fed. In this instance, enzyme activity was significantly lower in those quails fed the TRF mixture. Apparently, the level of dietary fat does not affect cholesterol metabolism in Japanese quails, even when high levels of TRF are included in the diet.

TABLE 4. Composition of Quail Diets.

Ingredient[a]	Dietary Fat (%)			
	0.75	1.50	3.00	6.00
Corn starch	510	502.5	487.5	457.5
Casein	220	220	220	220
Whey powder	30	30	30	30
Sucrose	100	100	100	100
Corn oil	5.5	5.5	5.5	5.5
Canola oil	1.0	4.75	12.25	27.25
Coconut oil	1.0	4.75	12.25	27.25
α-Cellulose	80	80	80	80
Guar gum	10	10	10	10
Mineral mixture[b]	40	40	40	40
Vitamin mixture[b]	2.5	2.5	2.5	2.5

[a]All data expressed in g/kg.
[b]As recommended by American Institute of Nutrition.
Source: Adapted from Hood (1995).

TABLE 5. Effect of Feeding Dietary Fat and Tocotrienol-Rich Fraction (TRF) for 3 Weeks on Cholesterol Metabolism of Japanese Quail.

Dietary Fat (%)	TRF[a]	Blood Cholesterol (mM)		Liver Cholesterol (mg/liver)	HMGCoA Reductase Activity
		Total	LDL		
0.75	+	3.92 ± 0.26	1.14 ± 0.13	8.6 ± 0.9	266 ± 45
0.75	−	4.02 ± 0.29	1.28 ± 0.16	10.8 ± 1.4	268 ± 32
1.50	+	4.19 ± 0.29	0.98 ± 0.12	9.4 ± 0.6	172 ± 27
1.50	−	3.59 ± 0.34	1.00 ± 0.13	9.1 ± 1.6	292 ± 62
3.0	+	4.15 ± 0.35	1.16 ± 0.20	9.8 ± 1.5	249 ± 42
3.0	−	4.59 ± 0.32	1.20 ± 0.26	10.1 ± 2.3	255 ± 53
6.0	+	4.20 ± 0.33	1.03 ± 0.18	8.5 ± 2.6	326 ± 65
6.0	−	4.14 ± 0.34	1.10 ± 0.15	8.9 ± 1.6	310 ± 73

[a]2 mg of TRF was administered daily with 50 μL of canola oil. Control birds received an equivalent amount of α-tocopherol (0.75 mg) in 50 μL of canola oil.
Mean ± standard error of the mean.
Source: Adapted from Hood (1995).

MECHANISM OF ACTION

The mechanism for inhibition of cholesterol biosynthesis by tocotrienols appears to involve post-transcriptional suppression of HMGCoA reductase (Parker et al., 1993) and is different from that of known inhibitors (e.g., lovastatin) of cholesterol synthesis. Pearce et al. (1992) demonstrated the importance of the tocotrienol side-chain unsaturation in their studies on structure-activity relationships regulating cholesterol synthesis. They concluded that tocotrienols lacking 5-methyl substitution (e.g., γ-tocotrienol and δ-tocotrienol) are more potent than α-tocotrienol in suppressing HMGCoA reductase activity.

Pearce and colleagues (1994) prepared the benzopyran and tetrahydronaphthalene analogues of tocotrienols to investigate their antioxidant and hypocholesterolemic properties. They concluded that the farnesyl side chain and hydroxy/methyl substitution pattern of γ-tocotrienol deliver a high level of HMGCoA reductase suppression, which is unsurpassed by synthetic analogues. Of the 42 analogues of tocotrienols they synthesized and fed to chickens, only 2-desmethyltocotrienol, δ-bromotocotrienol, and a tetrahydronaphthalene derivative exhibited a greater degree of LDL-cholesterol lowering than the natural tocotrienols. Even though tocotrienols exhibit hypocholesterolemic activity and antioxidant properties, Pearce and colleagues suggest that certain synthetic analogues of tocotrienols may prove valuable as antiatherosclerotic compounds.

The isoprenoid constituents of the diet, including tocotrienols, have been suggested to have a growth-suppressive action through interaction with the mevalonate biosynthetic pathway (Elson, 1996), which in the case of tocotrienols leads to a decrease in the mass and activity of HMGCoA reductase. This could limit the availability of cholesterol for membrane formation in proliferating tissues and might also interfere with signal transduction pathways involved in cellular proliferation (Kothapalli et al., 1993). A wide-ranging review by Elson discusses a number of issues relating to the suppression of HMGCoA reductase activity and cholesterol-lowering action of dietary isoprenoids (Elson et al., 1995).

DOSE-DEPENDENT INHIBITION OF HMGCoA REDUCTASE

The concentration-dependent impact of γ-tocotrienol on serum cholesterol can be traced to the posttranscriptional down-regulation of HMGCoA reductase activity (Parker et al., 1993). The action of the tocotrienols resembles that of the nonsterol component, believed to be farnesol (Correll et al., 1994). Tocols differ in their mevalonate-suppressive potency; *d*-γ and *d*-δ-tocotrienol are fourfold more potent than *d*-α-tocotrienol in suppressing cholesterol synthesis in rat hepatocytes, whereas dietary *d*-α-tocopherol increases avian hepatic HMGCoA reductase activity (Qureshi et al., 1996). In isolated hepatocytes, tocotrienols produce a dose-dependent suppression of HMGCoA reductase activity, whereas, in general, animal studies show that the impact of tocotrienols on HMGCoA reductase activity and blood cholesterol reaches a plateau (Qureshi et al., 1996b; Correll et al., 1994) and then wanes as the dose is further increased (Khor et al., 1995a, 1995b). There are conflicting reports of the impact of Palmvitee (see Table 3, Qureshi et al., 1996a), the commercial source of γ-tocotrienol, on HMGCoA reductase activity and serum cholesterol concentrations. These conflicting reports initiated studies (Qureshi et al., 1996b; Khor et al., 1995a, 1995b) on the impact of α-tocopherol on the cholesterol-suppressive action of γ-tocotrienol.

Qureshi et al. (1996b) fed groups of White Leghorn chickens either a control or 1 of 6 experimental diets for 26 days. The control diet contained 21 nmol of α-tocopherol per gram, and the experimental diets provided 141 nmol of α-tocopherol/γ-tocotrienol mixture in varying proportions. The α-tocopherol and γ-tocotrienol concentrations in the 6 diets ranged from 141 to 21 and 0 to 120 nmol/g, respectively. These researchers report that including α-tocopherol in the tocol mixtures containing adequate γ-tocotrienol to suppress HMGCoA reductase activity resulted in an alteration of the tocotrienol action. Based on an analysis of 10 other studies utilizing Palmvitee preparations, they suggest that effective preparations consist of 15 to 20% α-tocopherol and approximately 60% γ- and δ-tocotrienol, whereas less effective preparations consist of greater than 30% α-tocopherol and less than 45% γ- and δ-tocotrienol.

Khor's group (1995a, 1995b) used male guinea pigs to test the effect of different dosages of tocotrienols from palm oil fatty acid distillate (43% α-tocotrienol, 50% γ-tocotrienol, and 7% δ-tocotrienol). The experimental groups were given different dosages of tocotrienols dissolved in vitamin E-free palm olein triacylglycerols, namely 5, 8, 10, 15 and 50 mg/day for 6 days. Two additional groups were given 5 and 50 mg of α-tocopherol. The control guinea pigs were given vitamin E-free palm oil triacylglycerols for the same time period. Treated guinea pigs had only α-tocopherol and no tocotrienols in their serum, and it was noted that serum α-tocopherol concentrations increased with increasing dosages of administered tocotrienols. On the other hand, the livers of treated animals contained both α-tocopherol and α-, γ-, and δ-tocotrienols in a similar proportion to that administered in the vitamin E-free palm olein triacylglycerols.

Liver HMGCoA reductase activity of guinea pigs that were given different dosages of tocotrienols and tocopherols are shown in Table 6 (Khor et al., 1995b). When compared to animals on the control diet, they found that lower dosages (5 and 8 mg) of tocotrienols had a strong inhibitory effect on HMG-CoA reductase activity, whereas higher dosages (10 or 50 mg) of tocotrienols had a lesser inhibitory effect. When a high dosage of tocopherols (50 mg/day) was given, there was a significant stimulatory effect on HMGCoA reductase activity. Khor noted that there was a significant conversion of tocotrienols to α-tocopherol in the body because both liver and serum concentrations of α-tocopherol increased significantly in guinea pigs treated with tocotrienols.

TABLE 6. Liver HMGCoA Reductase Activity of Guinea Pigs Treated Intraperitoneally with Different Dosages of Pure Tocotrienols (T3) and Tocopherols (T) for 6 Consecutive Days.

Groups	Treatments (mg/d for 6 days)	HMGCoA Reductase $\pm$ SEM	% Activity
Control	POTG (n = 11)	14.5 ± 0.6^a	100
T3-treated	5mg (n = 5)	7.3 ± 0.6^b	50
	8mg (n = 4)	10.0 ± 0.7^b	69
	10mg (n = 7)	12.5 ± 1.2^b	86
	15mg (n = 5)	10.0 ± 0.6^b	69
	50mg (n = 6)	10.4 ± 1.4	72
T-treated	5mg (n = 5)	12.1 ± 0.9	83
	50mg (n = 5)	27.1 ± 3.5^b	187

n = number of animals.
POTG = vitamin E-free palm oil triacylglycerols.
Means with different superscripts are significantly different ($p < 0.05$).

These two findings led Khor's group to conclude that the lesser inhibitory effect on HMGCoA reductase activity by tocotrienols at higher dosages could be due to its conversion to tocopherols and the eventual accumulation of tocopherols in the tissue because high levels of α-tocopherol stimulate HMGCoA reductase activity in guinea pigs. In order to act as hypocholesterolemic agents, tocotrienols need to be supplied at low concentrations as pure compounds to animals that have been given minimal amounts of α-tocopherol.

TOCOTRIENOLS AND CANCER

A number of epidemiological studies have been concerned with relationships between diet and cancer and have provided evidence that the consumption of fruits and vegetables protects against various types of cancer (Wattenburg, 1992). This protective effect is generally attributed to the antioxidative capacities of vitamins C and E and b-carotene present in the food; however, minor components in fruits and vegetables, such as tocotrienols, isoprenoids, and flavonoids, may also be important.

Palm oil, unlike many other dietary oils, does not increase the yield of chemically induced mammary tumors in rats when fed at high levels in the diet (Sundram et al., 1989; Kritchevsky et al., 1992). The vitamin E fraction of palm oil, which is rich in tocotrienols, appears to be important in reducing certain tumors because palm oil stripped of the vitamin E fraction does increase tumor yields. Work in Carroll's laboratory (Nesaretram et al., 1992) showed that rats treated with the mammary carcinogen 7,12-dimethylbenz(a)-anthracene (DMBA) and fed vitamin E-free palm oil had more tumors than rats fed palm oil-containing tocols. Tocotrienols also caused a delay in the onset of subcutaneous lymphoma in HRS/J hairless mice by 2 to 4 weeks, and the life span of mice inoculated with transplanted tumor cells was increased by tocotrienols (Tan, 1992).

Tocotrienols inhibited proliferation and growth of both MDA-MB-435 and MCF-7 cells in culture much more effectively than did α-tocopherol (Guthrie et al., 1997a). When combined with tamoxifen, a drug widely used for treatment of breast cancer, tocotrienols inhibit growth of these cells more effectively than did tocotrienols or tamoxifen alone. There is an active ingredient in orange juice, probably the flavonoid hesperetin, that is able to reduce DMBA-induced mammary tumors (Guthrie et al., 1997b). Guthrie et al. (1997a) reported that a $1:1:1$ combination of δ-tocotrienol, hesperetin, and tamoxifen (IC_{50} 0.0005 μg/mL) was the most effective in inhibiting MCF-7 cells. Their results suggest that diets containing tocol-rich palm oil may reduce the risk of breast cancer, particularly when eaten with other plant foods containing flavonoids and tocols that may enhance the effectiveness of tamoxifen for treatment of breast cancer. Tocotrienols are effective inhibitors of both estrogen receptor-negative and -positive cells (So et al., 1997).

Elson's research team (Elson and Yu, 1994; Elson et al., 1995) plotted iso-prenoid-mediated suppression of hepatic HMGCoA activity and tumor growth and reported that the same isoprenoids (including tocotrienols) that suppressed HMGCoA activity also suppressed the growth of tumors. He reported a correlation of 0.98. His biological interpretation of this plot lends support to his contention that the potency with which an isoprenoid suppressed hepatic HMGCoA reductase activity accurately predicts that isoprenoid's antitumor action.

GRAIN AMARANTH TOCOLS

Amaranth was a sacred crop in the culture of the Aztec Indians. Grain amaranths together with maize and beans were once the staple diet of many Central American cultures. Traditionally, grain amaranth was popped and milled, which has been shown to improve the nutritional qualities of the grain (Singhal and Kulkarni, 1990). Grain amaranth is a pseudocereal that has many unusual nutritional and commercial properties, all of which have been summarized by Lehmann (1996). The tocols, four tocopherols and four tocotrienols, protect the seed oils in stored grain from rancidity and have been characterized (Lehmann et al., 1994; Qureshi et al., 1991a). Amaranth tocols typically contain 33% tocopherols, 61% β-tocotrienol, and 6% other tocotrienols, whereas popped amaranth contains 19% tocopherols, 47% β-tocotrienol and 34% other tocotrienols.

An international patent (Qureshi et al., 1996b) describes a method for the release of tocols from grains, including amaranth. The method uses both extrusion and inert (nonoxidizing) environments to achieve a 10- to 100-fold release of tocols. This process would be useful in food and dry applications, particularly for foods designed to lower LDL cholesterol. A more recent patent by Qureshi and colleagues (1993) indicates that, in addition to the eight vitamin E isomers, amaranth oil contains new and potent cholesterol inhibitors, namely, desmethyl-tocotrienol and didesmethyl-tocotrienol.

Chickens have a cholesterol biosynthetic pathway that is similar to humans (Leveille ct al., 1975); hence, Qureshi et al. (1996a) chose 6-week-old female chickens to compare the effects of variously processed amaranth varieties, an oily fraction of amaranth, and a standard corn-soy ration on cholesterol biosynthesis. Serum total cholesterol and LDL-cholesterol were lowered by 10 to 30% and 7 to 70%, respectively, in birds fed amaranth containing diets. High-density lipoprotein cholesterol was not affected. The activities of HMG-CoA reductase were lowered by only 9% for popped, milled amaranth and its oil. Qureshi and colleagues suggest that the lack of marked inhibition of this enzyme points to the presence of some other potent cholesterol inhibitors (e.g., plant isoprenoids) apart from tocotrienols and squalene in amaranth. Their results indicate that processing and varietal types are important when assessing hypocholesterolemic activities of grains and oils.

Squalene, a precursor in cholesterol biosynthesis, was first reported in amaranth oil by Lyon and Becker (1987). This product is commercially important as a lubricant and skin penetrant and is extracted from shark and whale liver oil. Amaranth oil contains approximately 6% squalene.

Grain amaranth has a number of potential cholesterol-lowering agents, including dietary fiber, squalene, tocotrienols, isoprenoid compounds, and/or other unknown factors. Danz and Lupton (1992) and Chaturvedi et al. (1993) attributed the hyprocholesterolemic effect of dietary amaranth to the content of squalene and dietary fiber. Others (Qureshi et al., 1993) have claimed that tocotrienols in amaranth oil are the inhibitors of cholesterol synthesis. Budin et al. (1996) have done an extensive investigation on the compositional properties of the seeds and oil of eight Amaranthus species and concluded that the hypocholesterolemic effects of dietary amaranth are apparently due to substances other than β-glucans or tocotrienols.

In a nutritional trial with Japanese quail that were given a balanced diet containing 60% grain amaranth, the addition of 0.2% squalene to the diet caused a 115% reduction in the activity of HMGCoA reductase; however, this was not reflected in an altered blood cholesterol profile (Hood, unpublished results). A similar observation was made when Hood used rats as the test animal.

CONCLUSION

This chapter has briefly reviewed the metabolic effects of tocotrienols, particularly those derived from palm oil, and amaranth, on cholesterol metabolism and tumor cells. There have been conflicting reports on the impact of tocotrienols on serum cholesterol concentration. The variation in experimental results can be explained by the proportion of α-tocopherol and tocotrienol in the dietary tocol mixtures, since α-tocopherol stimulates and tocotrienols suppress HMGCoA reductase activity. For tocotrienols to act as hypocholesterolemic agents, they need to be fed at low concentrations and as pure compounds to animals that have been given minimal amounts of α-tocopherol. High doses of dietary tocotrienols will cause a significant conversion of tocotrienol to α-tocopherol, which, in turn, results in a lower inhibitory effect by tocotrienol on HMGCoA reductase.

REFERENCES

Budin, J. T., Breene, W. M., and Putnam, D. H., 1996. Some compositional properties of seeds and oils of eight *Amaranthus* species, *J. American Oil Chemists' Society.* 73:475–481.

Burton, G. W., and Ingold, K. U., 1989. Vitamin E as an *in vitro* and *in vivo* antioxidant, *Annuals of New York Academy of Sciences.* 570:7–22.

Caroll, K. K., Guthrie, N., Nesaretnam, K., Gapor, A., and Chambers A. F. 1995. Anticancer properties of tocotrienols from palm oil in *Nutrition, Lipids, Health and Disease.* Champaign: AOCS Press, pp. 117–121.

Chaturvedi, A., Sarojini, G., and Devi, N. L., 1993. Hypocholesterolemic effects of amaranth seed (*Amaranthus esculantus*), *Plant Foods in Human Nutrition.* 44:63–70.

Correll, G. C., Ng, L., and Edwards, P. A., 1994. Identification of farnesol as the non-sterol derivative of mevalonic acid required for the accelerated degradation of 3-hydroxy-3-methylglutaryl coenzyme A reductase, *J. Biological Chemistry.* 269:17390–17393.

Danz, R. A., and Lupton, J. R., 1992. Physiological effects of dietary amaranth (*Amaranthus cruentus*) on rats, *Cereal Foods World.* 37:489–94.

Donaldson, W. E., 1982. Atherosclerosis in cholesterol-fed Japanese quail: evidence for amelioration by dietary vitamin E, *Poultry Science.* 61:2097–2104.

Elson, C. E., 1992. Tropical oils: nutritional and scientific issues, *Critical Reviews in Food Science and Nutrition.* 31:79–102.

Elson, C. E., 1996. Dietary fats, lipids, and tumorigenesis: new horizons in basic research, *Advances in Experimental Medicine and Biology.* 399:71–86.

Elson, C. E., and Yu, S. G., 1994. The chemoprevention of cancer by mevalonate-derived constituents of fruits and vegetables, *Journal of Nutrition.* 124:607–614.

Elson, C. E., Yu, S. G., and Quershi, A. A., 1995. The cholesterol- and tumor-suppressive actions of palm oil isoprenoids in *Nutrition, Lipids, Health and Disease.* Champaign: AOCS Press, pp. 109–116.

Evans, A. J., Hood, R. L., Oakenfull, D. G., and Sidhu, G. S., 1992. Relationship between structure and function of dietary fiber. A comparative study of effects of three galactomannans on cholesterol metabolism in the rat, *British Journal of Nutrition.* 68:217–229.

Goh, S. H., Choo, Y. M., and Ong, A. S. H., 1985. Minor constituents of palm oil, *Journal of American Oil Chemists' Society.* 62:237–240.

Guthrie, N., Gapor, A., Chambers, A. F., and Carroll, K. K., 1997a. Palm oil tocotrienols and plant flavonoids act synergistically with each other and with tamoxifen in inhibiting proliferation and growth of estrogen receptor-negative MDA-MB-435 and -positive MCF-7 human breast cancer cells in culture, *Asia Pacific J. Clin. Nutr.* 6:41–45.

Guthrie, N., Gapor, A., Chambers, A. F., and Carroll, K. K., 1997b. Inhibition of proliferation of estrogen receptor-negative MDA-MB-435 and -positive MCF-7 human breast cancer cells by palm oil tocotrienols and tamoxifen and in combination, *Journal of Nutrition.* 127(3):5445–5485.

Hood, R. L., 1990. Effect of diet and substrate on the *in vitro* measurement of cholesterol and fatty acid synthesis in hepatic tissue of Japanese quail, *Poultry Science.* 69:647–651.

Hood, R. L., 1991. Effect of dietary fats on hepatic cholesterol synthesis in Japanese quail, *Poultry Science.* 70:1848–1850.

Hood, R. L., 1995. Tocotrienols and Cholesterol Metabolism in *Nutrition, Lipids, Health and Disease.* Champaign: AOCS Press, pp. 96–103.

Hood, R. L., and Sidhu, G. S., 1992. Effect of guar gum and tocotrienols on cholesterol metabolism of Japanese quail, *Nutrition Research.* 12:117S–127S.

Jenkins, D. J. A., Leeds, A. R., Newton, C., and Cummings, J. H., 1975. Effect of pectin, guar gum and wheat fiber on serum cholesterol, *Lancet.* 1:1116–1117.

Khor, H. T., Chieng, D. Y., and Ong, K. K., 1995a. Tocotrienols inhibit HMGCoA reductase activity in the guinea pig, *Nutrition Research.* 15:537–544.

Khor, H. T., Chieng, D. Y., and Ong, K. K., 1995b. Tocotrienols — A dose-dependent inhibitor for HMGCoA reductase in *Nutrition, Lipids, Health and Disease.* Champaign: AOCS Press. pp. 104–108.

Kothapalli, R., Guthrie, N., Chambers, A. F. and Carroll, K. K., 1993. Farnesylamine: an inhibitor of farnesylation and growth of rat-transformed cells, *Lipids.* 28:969–973.

Kritchevsky, D., Weber M. M., and Klurfeld D. M., 1992. Influence of different fats (soybean oil, palm olein, or hydrogenated soybean oil) on chemically-induced mammary tumors in rats, *Nutrition Research.* 12:175s–179s.

LaRosa, J. C., Hunninghake, D., Bush, D., Criqui, M. H., Getz, G. S., Gotto, A. M., Grundy, S. M., Rakita, L., Robertson, R. M., Weisfeldt, M. L., and Cleeman, J. I. 1990. The cholesterol facts. A summary of the evidence relating dietary fats, serum cholesterol and coronary heart disease. A joint statement by the American Heart Association and the National Heart, Lung, and Blood Institute. The Task Force on Cholesterol Issues, American Heart Association, *Circulation.* 81:1721–1733.

Lehmann, J. W., 1996. Case history of grain amaranth as an alternative crop, *Cereal Foods World.* 41:399–410.

Lehmann, J. W., Putnam, D. H., and Qureshi, A. A., 1994. Vitamin E isomers in grain amaranths (*Amaranthus* spp.), *Lipids.* 29:177–181.

Leveille, G. A., Romsos, D. R., Yeh, Y. Y., and OíHea, E. K., 1975. Lipid biosynthesis in chicks. A consideration of site of synthesis, influence of diet and possible regulatory mechanisms, *Poultry Science.* 54:1075–93.

Lyon, C. K., and Becker, R., 1987. Extraction and refining of oil from amaranth seed, *J. American Oil Chemists' Society.* 64: 233–236.

Nesaretnam, K., Khor, H. T., Ganeson, J., Chong, Y. H., Sundram, K., and Gapor, A. 1992. The effect of vitamin E tocotrienols from palm oil on chemically-induced mammary carcinogenesis in female rats, *Nutrition Research.* 12:63–75.

Packer, L., 1995 Nutrition and biochemistry of the lipophilic antioxidants vitamin E and carotenoids in *Nutrition, Lipids, Health and Disease.* Champaign: AOCS Press, pp. 8–385.

Palca, J., 1990, Getting to the heart of the cholesterol debate, *Science.* 247:1170–1173.

Parker, R. A., Pearce, B. C., Clark, R. W., Gordon, D. A., and Wright, J. J., 1993. Tocotrienols regulate cholesterol production in mammalian cells by post-transcriptional suppression of 3-hydroxy-3-methyl glutaryl coenzyme A reductase, *J. Biological Chemistry.* 268:11230–11238.

Pearce, B. C., Parker, R. A., Deason, M. E., Dischino, D. D., Gillespie, E., Qureshi, A. A., Volk, K., and Wright, J. J. K., 1994. Inhibitors of cholesterol biosynthesis 2. Hypocholesterolemic and antioxidant activities of benzopyran and tetrahydronaphthalene analogues of the tocotrienols, *Journal of Medical Chemistry.* 37:526–541.

Pearce, B. C., Parker, R. A., Deason, M. E., Qureshi, A. A., and Wright, J. J. K., 1992. Hypocholesterolemic activity of synthetic and natural tocotrienols, *Journal of Medical Chemistry.* 35:3595–3606.

Qureshi, A. A., Burger, W. C., Peterson, D. M., and Elson, C. E., 1986. The structure of an inhibitor of cholesterol biosynthesis isolated from barley. *Journal of Biochemistry.* 261:10544–10550.

Qureshi, A. A., Becker, K. W., Wells, D. M., and Lane, R. H., 1991a. Processes for recovering tocotrienols, tocopherols and tocotrienol-like compounds, Int. Patent WO 91/17985.

Qureshi, A. A., Bradlow, B. A., Brace, L., Manganello, J., Peterson, D. M., Pearce, B. C., Wright, J. J. K., Gapor, A., and Elson, C. E., 1995. Response of hypercholesterolemic subjects to administration of tocotrienols, *Lipids.* 30:1171–1177.

Qureshi, A. A., Lane, R., and Salsers, A. W., 1993. U. S. Patent 91 U. S. 796486.

Qureshi, A. A., Lehmann, J. W., and Peterson, D. M., 1996a. Amaranth and its oil inhibit cholesterol biosynthesis in 6-week-old female chickens, *J. Nutrition.* 126:1972–78.

Qureshi, A. A., Pearce, B. C., Gapor, A., Nor, A., Peterson, D. M., and Elson, C. E. 1996b. Dietary alpha-tocopherol attenuates the impact of gamma-tocotrienol on hepatic 3-hydroxy-3-methyglutaryl coenzyme A reductase activity in chickens, *Journal of Nutrition.* 126:389–394.

Qureshi, A. A., Peterson, D. M., Elson, C. E., Mangels, A. R., and Din, Z. Z., 1989. Stimulation of avian cholesterol metabolism by α-tocopherol, *Nutrition Reports International.* 40:993–1001.

Qureshi, A. A., Qureshi, N., Hasler-Rapacz, J., Weber, F. E., Chaudhary, V., Crenshaw, T. D., Gapor, A., Ong, A. S. H., Chong, Y. H., Peterson, D., and Rapacz, J. 1991b. Dietary tocotrienols reduce concentrations of plasma cholesterol, apolipoprotein B, thromboxane B_2, and platelet factor 4 in pigs with inherited hyperlipidemias, *American Journal of Clinical Nutrition.* 53:1042S–1046S.

Qureshi, A. A., Qureshi, N., Wright, J. J. K., Shen, Z., Kramer, G., Gapor, A., Chong, Y. H., De Witt, G., Ong, A. S. H., Peterson, D. M., and Bradlow, B. A., 1991c. Lowering of serum cholesterol in hypercholesterolemic humans by tocotrienols (palmvitee), *American Journal of Clinical Nutrition.* 53:1021–1026S.

Serbinova, Z., Kagan, V., Han, D., and Packer L., 1992. Free radical recycling and intramembrane mobility in the antioxidant properties of alpha-tocopherol and alpha-tocotrienol, *Free Radicals in Biology and Medicine.* 10:263–275.

Shih, J. C., 1983. Atherosclerosis in Japanese quail and the effect of lipoic acid, *Federation Proceedings.* 42:2494–2497.

Singhal, R. S., and Kulkarni, P. R., 1990. Effect of puffing on oil characteristics of amaranth (Rajeera) seeds, *J. American Oil Chemists' Society.* 67:952–954.

So, F. V., Guthrie, N., Chambers, A. F., Moussa, M., and Carroll, K. K., 1997. Inhibition of human breast cancer cell proliferation and delay of mammary tumorigenesis by flavonoids and citrus juice, *Nutrition and Cancer.* 26:167–181.

Sundram, K., Khor, H. T., Ong, A. S. H., and Pathmanathan, R., 1989. Effects of dietary palm oils on mammary carcinogenesis in female rats induced by 7, 12-dimethylbenz(a)anthracene, *Cancer Research.* 49:1447–1451.

Tan, B., 1992. Anitumor effects of palm carotenes and tocotrienols in HRS/J hairless female mice, *Nutrition Research.* 12:163s–173s.

Tan, D. T. S., Khor, H. T., Low, W. H., Ali, A., and Gapor, A., 1991. Effect of a palm-oil-vitamin E concentrate on the serum and lipoprotein lipids of humans, *American Journal of Clinical Nutrition.* 53:1027S–1030S.

Tomeo, A. C., Geller, M., Watkins, T. R., Gapor, A., and Bierenbaum, M. L., 1995. Antioxidant effects of tocotrienols in patients with hyperlipidemia and carotid stenosis, *Lipids*. 30:1179–1183.

Wahlqvist, M. L., Krivokucα-Bogetic, Z., Lo, C. H., Hage, B., Smith, R., and Lukito, W., 1992. Differential serum responses to tocopherols and tocotrienols during vitamin E supplementation in hypercholesterolaemic individuals without change in coronary risk factors, *Nutrition Research*. 12:S181–S201.

Wattenburg, L. W., 1992. Inhibition of carcinogenesis by minor dietary constituents, *Cancer Research*. 52:2085s–2091s.

Phytochemical Interactions: β-Carotene, Tocopherol and Ascorbic Acid

STANLEY T. OMAYE
PENG ZHANG

INTRODUCTION

PHYTOCHEMICALS, the naturally occurring components of foods, some of which have defined nutrient functions, others whose functions are yet to be found, may have significant roles in health as part of a varied diet (Wardlaw and Insel, 1996; Bloch and Thomson, 1995; Reardon, 1995). These compounds can be obtained through the consumption of a diet from a normal food supply, which includes fruits, vegetables, grains, legumes, and seeds. Some phytochemicals may be found in rich deposits of specific foods, such as soy, wheat germ, green tea, colored vegetables, and fruits. Relative to health issues and broadly speaking, phytochemicals may be divided into health-promoting compounds and toxicants. Whether a compound falls into either group is a function of intake, i.e., dose makes the poison or the higher the intake, the more likely a toxic response may occur (Hathcock, 1993; Klaassen, 1996). Because scientific evidence is growing to support roles of many phytochemicals as health promoters or as adjuncts in preventing disease, there has been popular interest for increasing our phytochemical intake from sources rich in such compounds. Increased intakes of phytochemical sources can occur in a number of ways (Hathcock, 1995): (1) by increasing the amount in the diet of a plant food containing the phytochemical of interest, (2) through conventional plant breeding to increase the concentration of one or more target chemicals in a plant used in food, (3) through biotechnological techniques that genetically increase the concentration of one or more chemical in a plant used as food, (4) by processing, e.g., extraction and milling, to selectively enhance the concentration of one or more chemicals in a plant product as food, and (5) by adding concentrated extracts of natural or synthetic phytochemicals to food to enhance their concentrations in finished products. Enhancement of food phytochemicals through genetic engineering is already available with the growth of ascorbic acid-enriched citrus fruits, high-phytochemical broccoflower, and baked products made with fiber-enriched materials.

β-Carotene, vitamin E, and ascorbic acid are the chemicals most recognized as phytochemicals by the consumer; collectively, we know more about them than we know about any other phytochemical components of the diet. Although, technically, tocopherol (vitamin E) and ascorbic acid (vitamin C) are defined as nutrients, they can serve as prototypes for many phytochemicals. Thus, a better understanding of the physiological properties and interaction of β-carotene, vitamin E, and ascorbic acid will enable us to pursue development of an understanding of other phytochemicals, e.g., polyphenols, etc., (DeWalley et al., 1990; Jacob, 1995).

All three compounds, as well as many other phytochemicals, have been labeled as antioxidants (DeWalley et al., 1990). The physiological functions of antioxidants may be divided into three categories: preventive antioxidants, chain-breaking antioxidants, and repair and *de novo* compounds (Niki, 1991; Burton and Ingold, 1989; Sies, 1985; Chen et al., 1993). Preventive antioxidants are those compounds that reduce the rate of initiation of the free radical chain reaction, e.g., the selenium-dependent enzyme glutathione peroxidase (Reddy and Omaye, 1987). Chain-breaking antioxidants interact rapidly with the radicals after the chain reaction is initiated, converting radicals to stable entities and inhibiting the propagation phase, e.g., vitamins C and E. Repair and *de novo* compounds include enzymes that directly restore altered molecules to their original state or degrade them to nonfunctional compounds (catabolic reactions). β-Carotene can be both a preventive and chain-breaking antioxidant (Krinsky, 1989; Tappel, 1993; Chen et al., 1993).

BIOLOGICAL AND BIOCHEMICAL SIGNIFICANCE OF VITAMIN E

Tocopherols are four homologues that, with the tocotrienol counterparts, make up the majority of a class of compounds referred to as vitamin E. The tocopherol series has the phytyl side chain R_3, whereas, in the tocotrienols, the side chains have double bonds at $3'$, $7'$, and $11'$ positions. Vitamin E compounds are only synthesized by plants and are important dietary nutrients for humans and animals (Machlin, 1980; Packer and Fuchs, 1993; Kamal-Eldin and Appelqvist, 1996; Sokol, 1996). The oil seeds, leaves, and green parts of plants are rich in tocopherol, and α-tocopherol is mainly found in the chloroplast of plant cells and other homologues dispersed outside the organelle. Tocotrienols are found in the bran and germ fractions of seeds and cereals and not in the green parts of the plant. With the exception of palm oil, overall, tocotrienols are less widely found in nature than tocopherols. Table 1 is a list of some common rich sources of tocopherols and tocotrienols. Polyunsaturated oils are usually rich sources of vitamin E compounds. Together, these compounds play a concerted role in protecting the plant cell and lipids from autoxidation. Their presence in food will increase storage life and maintain food wholesomeness; therefore they are an important additive in the food industry (Kubow, 1992, 1993).

ANTIOXIDANT ROLE

Due to the ability to donate a phenolic hydrogen of the chromanol ring at position 6 to lipid-free radicals, tocopherols and tocotrienols exhibited antioxidant activity. Antioxidant activity is believed to be vitamin E's major biochemical function. Ester or ether groups at the 6 position of the chromanol ring eliminate antioxidant activity (Machlin, 1980; Packer and Fuchs, 1993). It is extensively recognized that the *in vivo* relative antioxidant activity of tocopherols is in the order of $\alpha > \beta > \gamma > \delta$, which also is the order of activity of the four tocopherols compared in a homogeneous solution in dichlorobenzene (Dillard, et al., 1983; Burton and Ingold, 1989; Kamal-Eldin and Appelqvist, 1996). The relative potency *in vitro,* however, can differ, depending on the environmental conditions such as relative concentrations of fats or oils, temperature, light, and the presence of other chemical species such as other antioxidants or pro-oxidants (Kamal-Eldin and Appelqvist, 1996). In suspensions containing fats, oils, or lipoprotein, the order of relative potency is the exact reverse (Olcott and Van Der Ven, 1968; Chow and Draper, 1974; Gottstein and Grosch, 1990; Kamal-Eldin and Appelqvist, 1996), illustrating the impact of various synergistic and antagonistic interactions. Such interaction potentially may have some biologic significance, that is, relative activity is not just a function of absolute chemical reactivities.

As an antioxidant, tocopherols and tocotrienols protect tissue lipids from free radical attack. Tappel (1962) proposed that vitamin E functions as an *in vivo* antioxidant. Vitamin E compounds are the primary chain-breaking antioxidants in membranes and lipoproteins, and, as such, reduce chemical species such as peroxyl, hydroxyl, and superoxide radicals and singlet oxygen. These species are normally formed by many cellular enzyme systems and exposure to various environmental agents (Reddy and Omaye, 1987; El-sayed et al., 1989; Omaye et al., 1991). Vitamin E reacts with lipid peroxyl radicals to form a relatively stable lipid hydroperoxide and vitamin E radical intermediate, which, in turn, may be recycled by other reductants, as illustrated in Figure 1. In membranes, high reactivities with various peroxyl radicals are important because vitamin E reacting with lipid peroxyl radicals to

TABLE 1. Selected Food Items Containing β-Carotene, Vitamin E, or Ascorbic Acid.

β-Carotene, μg/100g

Carrots	9800	Apricot	1500
Broccoli	700	Cantaloupe	3000
Lettuce	1200	Apple	26
Kale	4700	Cranberries	22
Beet	250	Grapefruit, pink	1310
Asparagus	450	Grapefruit, white	14
Cabbage	80	Kiwi fruit	43
		Lemon	3

Vitamin E, mg/100g

*Animal fats**		*Plant oils*	
Lard	0.6–1.3	Soybean	56–170
Butter	1.5	Cotton	30–80
Tallow	1.2–2.4	Maize	50–160
		Coconut	1–4
		Peanut	20–30
		Palm	30–70
		Safflower	25–50

Vitamin E, μg/g**

Grains		*Grains*	
Tocopherols		Tocotrienols	
Maize	35–70	Maize	40–85
Soybean	4–35	Soybean	trace
Cotton	5–35	Cotton	1–2
Oats	4–8	Oats	10–20
Milo	4–7	Milo	<1
Barley	11–14	Barley	20–30
Wheat	8–10	Wheat	2–3

Ascorbic Acid, mg/100g†

Fruits		*Vegetables*	
Apple	10–30	Broccoli	90–150
Banana	10	Cauliflower	60–80
Cherry	10	Kale	120–180
Grapefruit	40	Onion	10–30
Guava	300	Potato	10–30
Orange	50	Spinach	50–90
Rose hips	1000		

*Up to 90% as α-tocopherol.
**Mostly α- and γ-forms.
†Uncooked.

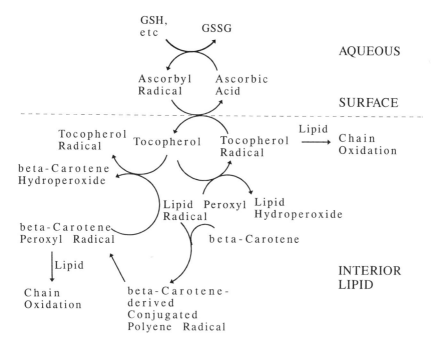

Figure 1 Schematic showing the proposed relationship between β-carotene, tocopherol, and ascorbic acid in the prevention of oxidative stress. Glutathione (GSH), oxidized glutathione, or peptide/protein sufhydryl groups (modified from Sies and Stahl, 1995; Niki et al., 1995).

produce a somewhat stable lipid hydroperoxide and the vitamin E radical interrupt the radical chain reaction, providing protection against lipid peroxidation.

Vitamin E has been often referred to as Nature's best chain-breaking antioxidant. Typically, one molecule of vitamin protects about 100 membrane phospholipids (Tappel, 1973; Kornbrust and Mavis, 1980; Sevanian, 1982; Liebler, 1993). This protection is due to vitamin E's ability to react 100 to 1,000 times faster with peroxyl radical than with phospholipid-bound polyunsaturated fatty acids (PUFAs) (Liebler, 1993; Machlin, 1980; Burton, 1994; Sokol, 1996).

PRO-OXIDANT ACTIVITY

Ideally, effective antioxidants should yield radicals that are unreactive toward stable molecules and that are limiting their reactions only to donation of hydrogens to radicals. However, antioxidants and their radicals often undergo other side reactions that may be classified as pro-oxidative. Whether antioxidant or pro-oxidant reaction occurs is determined by various factors such as their structure, concentration, temperature, etc.

There is a potential for vitamin E compounds to act as pro-oxidants, particularly the tocopheroxyl radical (Pokorny, 1987). When the concentration of tocopheroxyl radical is high enough, it is possible for a number of undesirable side reactions to occur, which in turn may initiate a chain reaction enhancing lipid peroxidation. In suspensions of low-density lipoproteins (LDLs) isolated from blood, vitamin E can accelerate the peroxidation of PUFAs under mild free radical conditions (Bowry et al., 1992; Bowry and Stocker, 1993; Ingold et al., 1993). It was suggested (Kamal-Eldin and Appelqvist, 1996) that, in situations where LDL peroxidation is initiated by reactions with various attacking aqueous radicals, vitamin E residing at or near the surface of the membranes or lipoprotein particles form vitamin E radicals. The location of vitamin E at different lipid levels of a lipoprotein particle is illustrated in Figure 2. Inasmuch because the LDL particle forces it to propagate the radical chain by its reaction with PUFA within the particle, such vitamin E radicals are not able to escape and propagate the radical chain with PUFA within the particle, particularly, if another reductant, such as ascorbic acid, is not available in the aqueous phase outside the LDL particle.

Food scientists have known about a similar situation in the food industry. Oxidation of food vegetable oils can occur when the vitamin content becomes greater than 2,000 ppm (Peers and Coxon, 1983; Terao and Matsushita, 1986). This can be accelerated by the presence of metal ions such as iron or copper and various pigments (Kamal-Eldin and Appelqvist, 1996).

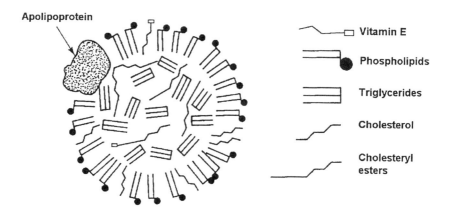

Figure 2 Vitamin E as a pro-oxidant in low-density lipoprotein (LDL) particle.

BIOLOGICAL AND BIOCHEMICAL SIGNIFICANCE OF ASCORBIC ACID

Like tocopherol, ascorbic acid has been defined as an essential nutrient. A diet devoid of ascorbic acid will result in scurvy, and the current recommended daily allowances reflect the level of an intake that will prevent this disease. Ascorbic acid is common in a variety of fresh fruits and vegetables, which are listed in Table 1. A particularly rich source is rose hips, with a concentration sometimes exceeding 1000 mg/g. With the exception of man, other primates, some bats and the guinea pig, most animals can synthesize ascorbic acid (Omaye et al., 1982).

In contrast to vitamin E, ascorbic acid is water soluble and is present in its deprotonated state under most physiologic conditions (Omaye et al., 1979; Sauberlich et al., 1982; Jacob et al., 1989). *In vivo,* most of the ascorbic acid is maintained in a reduced state by other endogenous reductants, and only some small amounts exist as dehydroascorbic acid or other oxidative products.

ANTIOXIDANT PROPERTIES

Strong evidence supports ascorbic acid's role as the most important antioxidant in extracelluar fluids. In addition, it has many cellular activities that may be directly or indirectly related to its antioxidant properties (Rose and Bode, 1993; Maritz, 1996; Barja, 1996). In studies with human plasma lipids, ascorbic acid was found to be far more effective in inhibiting lipid peroxidation initiated by peroxyl radical initiator than other components, such as protein thiols, urate, bilirubin, and α-tocopherol (Frei et al., 1989). Ascorbic acid can protect against lipid peroxidation by trapping the peroxyl radical in the aqueous phase before it can get into the lipid membrane or lipoprotein (Niki et al., 1995). Ascorbic acid has demonstrated to be an effective radical scavenger of superoxide, hydrogen peroxide, hypochlorite, hydroxyl radical, peroxyl radical, and singlet oxygen. Ascorbic acid can directly act as an antioxidant by reacting with aqueous peroxyl radicals or indirectly by restoring the antioxidant properties of other antioxidants, e.g., vitamin E in the lipid-soluble phase. The overall role of antioxidants is to control lipid peroxidation in membranes or lipid-containing particles (Jialal et al., 1990), which in turn maintains the fluidity and function of such lipid-containing materials. In addition, such deactivation of aqueous free radicals or oxidants by ascorbic acid lowers the attack on other important cellular components such as proteins and nuclear material (Wiseman, 1996; Halliwell, 1994).

PRO-OXIDANT EFFECT OF ASCORBIC ACID

The dual nature of ascorbic acid as an antioxidant and a pro-oxidant has lead to some confusion. Both actions are described in the literature (Herbert et al., 1996). As a pro-oxidant, ascorbic acid stimulates the peroxidation of lipids, proteins, enzymes, DNA, isolated mitochrondia, and tissue homogenates (Dasgupta and Zduneck, 1992; Stadman, 1991; Gordillo and Machado, 1991; Toyokuni and Sagripanti, 1992; Von Zglinicki et al., 1991; Lopez-Torres et al., 1992). Pro-oxidant effects are observed when dietary ascorbic acid results in tissue or blood plasma concentration in the micromolar range and when iron or copper is present in free forms (Halliwell and Gutteridge, 1989). The body goes to extreme efforts to keep divalent metal ions inactive and tied up in protein. In systems where metals are found free, ascorbic acid is used to maintain the metal in the reduced state and to supply electrons for free radical generation (Halliwell, 1994). However, when ascorbic acid reaches more physiological concentrations in the millimolar range, its effect on lipid peroxidation is mainly inhibitory. This was shown *in vivo* by feeding guinea pigs various levels of ascorbic acid and evaluating the amount of oxidative damage in their livers (Barja et al., 1994). Ascorbic acid stimulated liver peroxidation *in vitro,* in the presence of small amounts of iron in liver samples from guinea pigs that received a diet marginally deficient in vitamin C (33 mg ascorbic acid per kilogram) or when tissue reached micromolar concentrations of ascorbic acid. However, if the guinea pig was supplemented with higher dietary levels of ascorbic acid (660 mg or 13.2 g/kg) and liver vitamin concentrations reached the millimolar range, the rate of hepatic lipid peroxidation was strongly depressed even in the presence of iron.

For some individuals, high intakes of vitamin C may be counterindicated, particularly in those people with iron-overload problems (Halliwell, 1994; Herbert et al., 1996). This is because of the vitamin's ability to release transitional metals such as iron and copper from protein complexes and/or reduce these metals to their catalytic form (Chow, 1988). The concentration and subcellular distribution of ascorbic acid are important factors in determining the antioxidant or pro-oxidant function of ascorbic acid.

It is estimated that 10% of non-blacks and up to 30% of blacks have a gene disorder for iron overload (Herbert et al., 1996; Yip, 1994). This may be true for about 20% of American males (Gordeuk et al., 1992), where the iron may be available unbound either to transferrin or ferritin, thus exerting adverse catalytic reactions. Haber and Weiss (1934) recognized that free radical intermediates drove the Fenton reaction, i.e., the powerful oxidizing properties of a solution of ferrous salt and hydrogen peroxide (Ryan and Aust, 1993). Most of the $\bullet$OH generated *in vivo* comes from iron-dependent reduction of H_2O_2

(Gutteridge and Halliwell, 1990; Halliwell et al., 1991; Halliwell, 1994). As ferrous iron reduces H_2O_2 to generate $^•OH$ and ferric iron, supplementing individuals with vitamin C will promote the conversion of a ferric iron back to ferrous iron. This leads to a cycle where vitamin C provides a constant supply of reducing agents, turning a single incident of cycle or iron-dependent OH generation into a series of cycles of iron-dependent repetitive free radical generation. Depending on the circumstances, vitamin C can be a double-edged sword, where on one edge it is essential for health and acts as an antioxidant, and on the other edge it promoted pro-oxidant actions. Likewise, vitamin C enhances iron absorption in the gut and enhances the release of excess catalytic iron from harmless iron stores (Herbert et al., 1994; Lauffer, 1992). There have been reported cases where thalassemic patients with conditions of a heavy iron overload have died of vitamin C supplements (Herbert et al., 1994). Vitamin C supplements may have caused the rapid progression to death in the cardiomyopathy of hemochromatosis (McLaran et al., 1982). Others have noted that patients with high LDL cholesterol plus elevated serum ferritins have more than twice the coronary artery disease risk of those with lower serum ferritins and high LDL cholesterol (Salonen et al., 1992). *In vitro* (O'Connell et al., 1986; Herbert et al., 1994), iron-catalyzed lipid peroxidation was enhanced when vitamin C was added in physiological amounts. Thus, those afflicted with high LDL cholesterol and with moderately elevated body iron, such as those with hemochromatosis, may be placed at more risk for health disorders such as heart disease, cancer, and even aging if they are taking vitamin C supplements.

BIOLOGICAL AND BIOCHEMICAL SIGNIFICANCE OF β-CAROTENE

β-Carotene is one of 600 compounds identified as carotenoids (Wang, 1994; Olson and Krinsky, 1995). The only defined role of β-carotene in nutrition is as a precursor for the formation of vitamin A (Krinsky, 1993; Olson, 1996). In plants, carotenoids function in photoprotection or light collection (Demmig-Adams et al., 1996). Most carotenoids contain an extended system of conjugated double bonds, which are responsible for their antioxidant activity. The literature strongly provides evidence for β-carotene's role as an antioxidant and some indication that, under a specific situation, it can be causative in oxidation. Carotenoids usually have at least nine conjugated double bonds in their chemical structures before they meet the criterion for carotenoid efficacy as antioxidants.

Biological effects of carotenoids have been characterized either as functions, actions, or association (Olson, 1989; Krinsky, 1993). Carotenoid functions

include those as accessory pigments in photosynthetic organisms or protection against light sensitization (Cogdell and Frank, 1987; Mathews-Roth,
1987; Will and Scovel, 1989). In the case of animals on vitamin A-deficient diets, the metabolism of provitamin A carotenoids to retinol and retinoic acid are
viewed as a biological function of carotenoids (Olson, 1996). Actions have
been described as physiological or pharmacological responses to the administration of carotenoids (Olson, 1989; Krinsky, 1993). The role of carotenoids as
antioxidants and their ability to interact and quench free radicals is an example
(Liebler, 1993). Many effects attributed to carotenoids still lack sufficient evidence for confirmation and remain only statistical associations (Olson, 1989;
Krinsky, 1993). Research that targets development of a better understanding
about the relationship among structure, species, and observed or biological association of carotenoids could become fruitful because of the wide-reaching
potential of impacting human health.

   Adaptation to β-carotene in the diet of rodent occurs rapidly, that is,
both the uptake and the rate of conversion to vitamin A increase (Krinsky,
1993). So rapid is the conversion of β-carotene to vitamin A in rats that
such metabolism can confound the results of feeding β-carotene to rats.
Subsequently, investigators using rats have resorted to using pharmacological doses in order to produce significant increases in plasma and tissue levels of β-carotene. Other carotenoids, such as canthaxanthin, can raise
plasma and tissue carotenoid levels 3- to 6-fold greater than β-carotene
levels. Canthaxanthin and apocarotenal are two carotenoids that are struc-
turally related to β-carotene but lack vitamin A; their activities, yet, are reported to be as effective as β-carotene in quenching singlet oxygen and
trapping free radicals.

ANTIOXIDANT VERSUS PRO-OXIDANT PROPERTIES

   There is evidence that indicates that β-carotene is an important singlet oxygen and free radical scavenger (Burton and Ingold, 1984; Wolf, 1982; Liebler,
1993; Ozhogina and Kasaikina, 1995). Much of the evidence supporting the
antioxidant properties of β-carotene was performed using homogeneous solutions. Homogeneous solutions avoid the solubility problems, when both the
carotenoid and the lipids are dissolved in organic solvent. For reasons noted
above for vitamin E, one must be cautious with the interpretation of such
studies. Laboratories have found that the antioxidant activities of β-carotene
and carotenoids are related to the oxygen concentration (Burton and Ingold,
1984: Liebler and McClure, 1996). At 150 torr (composition of air), α-tocopherol is about 40- to 50-fold better than β-carotene as an antioxidant. However, at 15 torr (approximately the level of oxygen found in living tissues),

the difference in effectiveness decreases by 40%, that is, β-carotene has enhanced antioxidant activity at lower oxygen tension. At even higher oxygen tension (760 torr or 100% oxygen) any antioxidant activity of β-carotene appears to be offset by β-carotene acting as a pro-oxidant. If such results are extrapolated to *in vivo* situations, the potential harm versus benefits of β-carotenes are alarming. β-Carotene would best be useful when oxygen was less available (hypoxia) to ambient situations. However, β-carotene would be contraindicated in a situation of high oxygen tension, such as hyperoxia, oxygen therapy, and in reperfusion injury.

β-Carotene is used clinically to prevent photosensitized tissue damage in humans with erythropoietic porphyria (Mathews-Roth, 1986). β-Carotene may protect against photosensitized tissue injury by scavenging free radicals and by quenching singlet oxygen (Krinsky, 1993). As noted above, the β-carotene antioxidant chemistry displays a striking dependence on oxygen tension (Burton and Ingold, 1984; Samokyszyn and Marnett, 1987; Stratton et al., 1993; Liebler and McClure, 1996), whereas increasing oxygen tension β-carotene is readily autoxidized and may display pro-oxidant behavior. At 150 torr oxygen pressure, β-carotene acts as an antioxidant in a rat liver microsomal membrane, inhibiting lipid peroxidation rate (Palozza et al., 1995). When oxygen pressure was increased to 760 torr, β-carotene acts as a pro-oxidant in microsomal membranes (Palozza et al., 1995), similar to that observed for lipid solutions (Burton and Ingold, 1984) or in liposomes (Kennedy and Liebler, 1992; Bowen and Omaye, 1997). On the other hand, oral β-carotene supplementation in humans did not provide direct antioxidant protection to LDL particles, neither at moderate nor high doses of β-carotene and at 15 or 35 torr (Reaven, 1994).

We found that, at 150 torr, the enrichment of human LDL with β-carotene increased copper-initiated lipid peroxidation. Kinetic patterns of oxidation support the suggestion that β-carotene promotes autocatalysis (Bowen and Omaye, unpublished results). After an initial antioxidant action, β-carotene appears to participate in the autocatalytic chain reaction of fatty acid peroxidation in the propagation phase of LDL oxidation. LDL β-carotene concentration had a significant linear relationship with the rate of LDL oxidation [Figure 3(a)], exponential relationship with the final concentration of thiobarbituric acid reactive substances [Figure 3(b)], and a linear relationship with β-carotene consumed in 60 minutes [Figure 3(c)]. It seems that oxidation occurs despite high concentrations of α-tocopherol. Thus, increased amounts of β-carotene in LDL may overwhelm any protective effect exerted by α-tocopherol.

To conclude, the antioxidant efficacy of β-carotene depends on oxygen tension, β-carotene concentration, and whether incorporation of β-carotene occurs *in vivo* or *in vitro*.

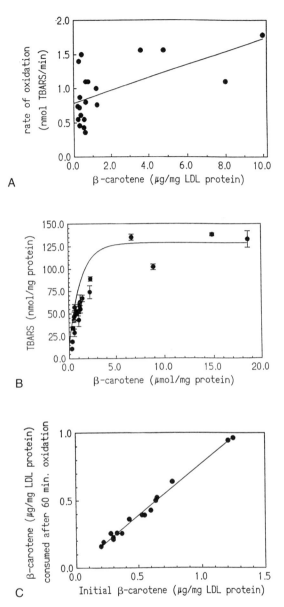

A

B

C

Figure 3 Relationship between low-density lipoprotein (LDL) oxidation and β-carotene (Bowen and Omaye, unpublished results): (a) LDL β-carotene concentration and rate of oxidation ($n = 22$; $r = 0.6102$; $p < 0.005$); (b) LDL β-carotene concentration and TBARS concentration ($n = 22$; $r = 0.8053$; $p < 0.005$); (c) Initial LDL β-carotene concentration and amount of β-carotene oxidized in 60 minutes ($r = 0.995$; $p < 0.001$). Thiobarbituric acid reactive substances (TBARS); (c) Initial LDL β-carotene concentration and amount of β-carotene oxidized in 60 minutes ($r = 0.995$; $p < 0.001$). Thiobarbituric acid reactive substances (TBARS).

ASCORBIC ACID INTERACTIONS WITH TOCOPHEROL

Interactions between various antioxidants would be expected to occur based on the order of reactivity of their oxidation-reduction potentials, as shown in Table 2 (Buettner, 1993; Olson, 1996). Any radical in a reaction identified toward the top of the table will react with a product in any reaction below it in the table. The actual reactivity may differ from prediction because of the dependence of such reactions as pH differences and substrate-product concentrations.

The functional interaction between tocopherol and ascorbic acid has long been recognized (Chow, 1991). The antioxidant function and requirement for vitamin E seem to be related to the status of ascorbic acid (Chen and Chang, 1979; Bendich et al., 1986). As a major water-soluble antioxidant, ascorbic acid may be involved in the regeneration or restoring of the antioxidant properties of vitamin E (Figure 1). After vitamin E has exerted its antioxidant function, it is converted to α-tocopheryl quinone, an oxidation product identified in tissues (Machlin, 1980: Chow, 1985). It has been demonstrated *in vitro,* or with chemical systems, that ascorbic acid is involved in the regeneration of vitamin E at the radical stage (tocopheroxy radical) prior to its irreversible conversion to the tocopheryl quinone. This recycling may not only occur at membranes but at the aqueous lipid interphase of lipoproteins (Kagan et al., 1992). Other reductants beside ascorbic acid, such as glutathione, may contribute toward the maintenance of reduced vitamin E (Chow, 1991). According to Niki (1995), ascorbic acid and tocopherol are located in different domains and interact at the interface between membrane and lipoprotein and water (Figure 1). Ascorbic acid is less effective when oxidation occurs deeper within the lipids. Spin-label techniques illustrated that the reduction of stable oxygen radicals by ascorbic acid became slower as the radical went

TABLE 2. Redox Potential of β-Carotene, Tocopherol, Ascorbic Acid, and Selected Other Antioxidants (Buettner, 1993; Olson, 1996).

Reaction	$E^{0'}$
$RS + e^- \rightarrow RS^-$	920
$Retinol^+ + e^- \rightarrow Retinol$	< 920
$\beta\text{-Carotene}^+ + e^- \rightarrow \beta\text{-Carotene}$	600
$\alpha\text{-Tocopheroxyl} + H^+ + e^- \rightarrow \alpha\text{-Tocopherol}$	500
$Ascorbate + H^+ + e^- \rightarrow Ascorbate^-$	280
$Dehydroascorbate + e^- \rightarrow Ascorbate^-$	−170
$\beta\text{-Carotene} + e^- \rightarrow \beta\text{-Carotene}^-$	−175
$RSSR + e^- \rightarrow RSSR^-$	−1500

deeper into the interior of the lipid material (Takahashi et al., 1989; Niki et al., 1995). Also, as noted before, ascorbic acid may react directly with free radicals and spare the oxidative degradation of vitamin E to tocoperoxyl radical. Ascorbic acid is capable of maintaining sulfhydryl compounds in a reduced state, participating in many redox reactions and scavenging singlet oxygen and free radicals.

β-CAROTENE INTERACTIONS WITH TOCOPHEROL

Little is known about the interactions between dietary vitamin E and β-carotene, particularly with how they might function together in protection against oxidative stress. Populations with low intakes of dietary fruits and vegetables, or perhaps low intakes of tocopherol and β-carotene, appear to be at greater risk for developing degenerative diseases (Gey, 1993; Heliovaara et al., 1994). However, the impact of other confounding factors, such as other phytochemicals in fruits and vegetables, remains to be determined. The relationship between the two compounds and their combined effect on oxidative stress may be quite complex. For example, it has been shown that supplementation of β-carotene decreased the plasma and hepatic levels of α-tocopherol in rats (Blakely et al., 1990) and only plasma levels in the chick (Mayne and Parker, 1986). Supplementation of rats with astaxanthin reduced the lysis of erythrocytes subjected to oxidative stress (Niki, 1991). Both astaxanthin or β-carotene supplements significantly lowered chick plasma vitamin E levels (Lim et al., 1992). The effect of lowering plasma and hepatic levels of α-tocopherol by β-carotene and other carotenoids may be explained by decreased α-tocopherol retention. β-Carotene could have altered absorption or binding of vitamin E in the liver because β-carotene and α-tocopherol may compete with each other for binding sites on lipoprotein molecules (Goodman et al., 1966; Pellet et al., 1994). Supportive evidence includes observations where humans fed 800 mg of α-tocopherol has decreased plasma carotenoids (Willett et al., 1983).

Conversely, chicks fed a vitamin E-deficient diet but given canthaxanthin supplements were reported to have higher tissue vitamin E levels compared to those animals who were maintained on a vitamin E-deficient diet alone (Mayne and Parker, 1989), suggesting that canthaxanthin protected vitamin E against oxidation. Recently, it was found that supplementation of chick diets with zeaxanthin or canthaxanthin had no effect on plasma or tissue levels of tocopherol; however, supplementation with β-carotene lowered plasma vitamin E by 50% (Woodall et al., 1996). Others have found that, when β-carotene and vitamin E were added to the diet of chicks using a 3 × 2 factorial arrangement of treatments, vitamin E protected against both heme protein oxidation and lipid peroxidation in various tissue extracts. Addition of β-carotene to the diet did not protect chicks significantly against either and substantially lowered retention of hepatic vitamin E (Leibovitz et al., 1990).

It has been suggested that α-tocopherol and β-carotene have complementary roles relative to the varying oxygen tension in biological membranes, since as noted above the antioxidant activity of β-carotene improves at physiological oxygen tensions. α-Tocopherol and β-carotene exert an additive effect in inhibiting radical-initiated lipid peroxidation of lipid extracts from rat liver microsomes using hexane solutions (Palozza and Krinsky, 1991). Other studies support a synergistically protective relationship between the two antioxidants (Leibovitz et al., 1990; Palozza and Krinsky, 1992).

The following hypothesis was offered to explain the synergistic interaction between α-tocopherol and β-carotene that results in both an increase of membrane resistance to oxidative stress and an increase of the loss of α-tocopherol (Palozza and Krinsky, 1992; Krinsky, 1993). At atmospheric oxygen concentration, a chain reaction initiating β-carotene peroxyl radical (β-COO$^\bullet$) can be formed. The additions of both antioxidants substantially facilitate the antioxidant character of β-carotenes by limiting the production or reactivity of β-COO$^\bullet$. The added tocopherol is consumed as it protects the β-carotene from autoxidation. Such protection is dose dependent and occurs only at high concentrations of tocopherol with a lack of synergism when levels of tocopherol were much lower than the levels of β-carotene (Palozza and Krinsky, 1992).

Alternatively, α-tocopherol could protect β-carotene by the recycling of an α-tocopherol radical; however, based on pulse radiolysis study, that is unlikely because of the lower ability of β-carotene to act as hydrogen or electron donor than α-tocopherol (Wilson, 1983).

ASCORBIC ACID AND OTHER CAROTENOID INTERACTIONS WITH β-CAROTENE

Based on the probable localization of β-carotene deep within the lipid bilayer of membrane or core of lipoproteins, it has been suggested that cooperative interaction between ascorbic acid and β-carotene is unlikely (Niki et al., 1995). β-Carotene has lower reactivity toward radicals than does α-tocopherol and acts as a weak antioxidant in solution. Because it is more lipophilic than α-tocopherol, it is assumed to be present at the interior of membranes or lipoproteins. This relationship between the lipid and aqueous layers is illustrated in Figure 1. It has been shown that the efficiency of radical scavenging by tocopherol decreases as the radical goes deeper into the interior of the membrane (Gonez-Fernandez et al., 1989) or lipoproteins (Takahashi et al., 1989). Both the water-soluble form of ascorbate and the lipid-soluble 6-O-palmitoylascorbic acid spared tocopherol but not β-carotene. Thus, there is no evidence to link ascorbic acid and β-carotene as synergistic protectors against oxidative damage.

In guinea pigs, plasma levels of retinol are enhanced by dietary ascorbic acid (Poovavia and Omaye, 1986). Of interest is the extent to which ascorbic acid may influence β-carotene or carotenoids during absorption; however, to our knowledge, no work has been reported. Carotenoids may compete with each other or with other lipid-soluble nutrients (Parker, 1996). Canthaxanthin or lycopene reduced plasma β-carotene levels in ferrets (White et al., 1993). A complex interaction between β-carotene and carotenoids has been observed in humans (Kostic et al., 1995). β-Carotene reduced the plasma lutein response, but lutein either enhanced or suppressed the plasma β-carotene response, depending on the individual.

CONCLUSION

Antioxidant protection involves a variety of chemical systems, i.e., endogenous compounds, including various enzymes and exogenous factors, of which many might be derived from the numerous phytochemicals found in our diet. It is apparent that, because of extensive interaction, both synergistic and antagonistic, that occurs *in vivo,* we can no longer assume that the outcome of their total action is equal to the sum of each antioxidant entity alone. For example, the intake of excessive amounts of β-carotene and perhaps other carotenoids may lead to reduced vitamin E levels rather than a synergistic response. Studies on a single antioxidant may be misleading if likely interactions are not considered. Increasing tissue concentration of an antioxidant at the expense of another may compromise the overall protection afforded to organisms by dietary antioxidants. The lack of information regarding such interactions raises concerns about making recommendations as to the indiscriminate use of antioxidants supplements. The use of antioxidant supplements upsets the concepts of balance, moderation, and variety, which are viewed by most nutritional scientists as the foundation for a healthy diet (RDA, 1989). It is important that research targets a better understanding of such interactions, particularly to determine when certain antioxidants spare some and when they may compromise others. There is a need for more research focusing on the effects of multiple combinations of antioxidants and phytochemicals. It is likely that other interactions with antioxidant compounds will be discovered in food; therefore, we must work toward multifactorial studies because this will add to understanding about the body's antioxidant defense system. The observation from the α-Tocopherol and β-Carotene Study and the β-Carotene and Retinol Efficacy Trial demonstrating a higher incidence of lung cancer among those who received β-carotene was unexpected and contrary to findings from epidemiologic studies. Prior to the initiation of these trials, most would have agreed that, based on epidemiologic, animal, and *in vitro* evidence, β-carotene might reduce the incidence of lung cancer and should be safe. With hindsight, it is easy to criticize and find judgment error; however, it is more important that we profit from the experience and consider what we can improve on in future studies. These trials remind us that no biologically active substance is safe at all dose levels for everyone. It reminds us that selection of a single food component or phytochemical is unlikely to reveal a "magic bullet." It is apparent that stresses, either environmental or diet related, can impact on whether unwanted side effects are produced by a nutrient recognized as safe. Study designs and clinical trial protocols must take into consideration the concept of the "total antioxidant pool" rather than placing emphasis on one individual antioxidant or phytochemical with antioxidant properties (Meyskens, 1990; Blakely et al., 1993). The formidable challenge for nutrition research is the design and implementation of experimental studies targeting the development of an understanding of the many antioxidant interactions.

REFERENCES

Barja, G. Ascorbic acid and aging. In: *Ascorbic Acid: Biochemistry and Biomedical Cell Biology, Subcellular Biochemistry Series.* (Harris, J. R., Ed.), Volume 25, Plenum Press, NY, pp. 157–188, 1996.

Barja, G., Lopiez-Torres, M., Perez-Campo, R., Rojas, C., Cadenas, S., Prat, J., and Pamplona, R. Dietary vitamin C decreases endogenous protein oxidative damage, malondialdehyde, and lipid peroxidation and maintains fatty acid unsaturation in guinea pig liver. *Free Rad. Biol. Med.* 17:105–115, 1994.

Bendich, A., Machlin, L. J., Scandurra, O., Burton, G. W., and Wayer, D. D. M. The antioxidant role of vitamin C. *Ad. Free Rad. Biol. Med.* 2:419–444, 1986.

Blakely, S. R., Grundel, E., Jenkins, M. Y., and Mitchell, G. V. Alterations in β-carotene and vitamin E status in rats fed β-carotene and excess vitamin A. *Nutr. Res.* 10:1035–1044, 1990.

Blakely, S. R., Mitchell, G. V., Jenkins, M. L. Y., and Grundel, E. Effects of β-carotene and related carotenoids on vitamin E. In: *Vitamin E in Health and Disease.* (Packer, L. and Fuchs, J., Eds.), Marcel Dekker, Inc., New York, NY, pp. 63–68, 1993.

Bloch, A. and Thomson, C. A. Position of the American Dietetic Association: Phytochemicals and functional foods. *J. Am. Diet Assoc.* 95:493–496, 1995.

Bowen, H. T. and Omaye, S. T. α-Tocopherol, β-carotene, and oxidative modification of human low density lipoprotein. In: *Lipid Peroxidation and Antioxidants.* (Baskin, S. I. and Salem, H. Eds.), Taylor & Francis, Washington, D.C., 1997.

Bowen, H. T. and Omaye, S. T. (Unpublished results), 1997.

Bowry, V. W., Ingold, K. U., and Stocker, R. Vitamin E in human low-density lipoprotein. *Biochem.* 288:341–344, 1992.

Bowry, V. W. and Stocker, R. Tocopherol-mediated peroxidation: The pro-oxidant effect of vitamin E on the radical-initiated oxidation of human low-density lipoprotein. *J. Am. Chem. Soc.* 115:6029–6044, 1993.

Buettner, G. A. The pecking order of free radicals and antioxidants: Lipid peroxidation, α-tocopherol, and ascorbate. *Arch. Biochem. Biophys.* 300:535–543, 1993.

Burton, G. W. Vitamin E: Molecular and biological function. *Proc. Nutr. Soc.* 53:251–262, 1994.

Burton, G. W. and Ingold, K. U. Autooxidation of biological molecules. 1. The antioxidant activity of vitamin E and related chain-breaking phenolic antioxidants *in vitro. J. Am. Chem. Soc.* 103:6472–6477, 1981.

Burton, G. W., and Ingold, K. U. β-Carotene: an unusual type of lipid antioxidant. *Science* 224:569–573, 1984.

Burton, G. W., and Ingold, K. U. Mechanisms of antioxidant action: Preventive and chain-breaking antioxidant action: Preventive and chain breaking antioxidants. In: *CRC Handbook of Free Radicals and Antioxidants in Biomedicine* (II), (Miquel, J., Quintanilha, A. T., and Weber, H., Eds.), CRC Press, Boca Raton, FL, pp. 29–44, 1989.

Chen, L. H. and Chang, H. M. Effects of high level of vitamin C on tissue antioxidant status of guinea pigs. *J. Int. Vit. Nutr. Res.* 49:87–91, 1979.

Chen, H., Pellett, L. J., Andersen, H. J., and Tappel, A. L. Protection by vitamin E, selenium, and β-carotene against oxidative damage in rat liver slices and homogenate. *Free Rad. Biol. Med.* 14:473–482, 1993.

Chow, C. K. Vitamin E and blood. *World Rev. Nutr. Diet* 45:133–166, 1985.

Chow, C. K. Interrelationships of cellular antioxidant defense systems. In: *Cellular Antioxidant Defense Mechanisms, Vol. II* (Chow, C. K., Ed.), CRC Press, Boca Roca, FL, pp. 217–237, 1988.

Chow, C. K. Vitamin E and oxidative stress. *Free Rad. Biol. Med.* 11:215–232, 1991.

Chow, C. K. and Draper, H. H. Oxidative stability and activity of the tocopheols in corn and soybean oils. *Int. J. Vit. Nutr. Res.* 44:396–403, 1974.

Cogdell, R. J. and Frank, H. A. How carotenoids function in photosynthetic bacteria. *Biochim. Biophys. Acta.* 895:63–79, 1987.

Dasgupta, A. and Zduneck, T. *In vitro* peroxidation of human serum catalyzed by cupric ion: Antioxidant rather than prooxidant role of ascorbate. *Life Science* 50:875–882, 1992.

Demmig-Adams, B., Gilmore, A. M., and Adams, W. W. III. *In vivo* functions of carotenoids in higher plants. *FASEB J.* 10:403–412, 1996.

DeWalley, C. V., Rankin, S. M., Hoult, J. R. S., Jessup, W., and Leake, D. S. Flavonoids inhibit the oxidative modification of low density lipoproteins by macrophages. *Biochem. Pharmac.* 39:1743–1750, 1990.

Dillard, C. J., Gavino, V. C., and Tappel, A. L. Relative antioxidant effectiveness of α-tocopherol and γ-tocopherol in iron-loaded rats. *J. Nutr.* 113:2226–2273, 1983.

Elsayed, N. M., Omaye, S. T., Klain, G. J., Inase, J. L., Dahlberg, E. T., and Korte, D. W., Jr. Mouse brain response to subcutaneous injection of the monofunctional sulfur mustard, butyl 2-chloroethyl sulfide (BCS). *Toxicology* 58:11–20, 1989.

Frei B., England, L., and Ames, B. N. Ascorbate is an outstanding antioxidant in human blood plasma. *Proc. Natl. Acad. Sci.* 86:6377–81, 1989.

Gey, K. F. Prospects for the prevention of free radical disease, regarding cancer and cardiovascular disease. *Brit. Med. Bull.* 49:679–699, 1993.

Gonez-Fernandez, J. C., Villalain, J., and Aranda, F. J. Localization of α-tocopherol in membranes. *Ann. NY Acad. Sci.* 570:109–120, 1989.

Goodman, D. S., Blomstrand, R., Werner, B., Huang, H. S., and Shiratori, T. The intestinal absorption and metabolism of vitamin A and β-carotene in man. *J. Clin. Invest.* 45:1615–1622, 1966.

Gordeuk, V., McLaren, C. E., Looker, A., and Brittenham, G. M. Evidence from NHANES II that the gene for heredity hemochromatosis is common. *Blood* 80:280, 1992.

Gordillo, E. and Machado, A. Implication of lysine residues in the loss of 6-phosphogluconate dehydrogenase activity in aging human erythrocytes. *Mech. Aging Dev.* 59:291–297, 1991.

Gottstein T., and Grosch, W. Model study of different antioxidant properties of α- and γ-tocopherol in fats. *Fat Sci. Technol.* 92:139–144, 1990.

Gutteridge, J. M. C. and Halliwell, B. Iron and oxygen: a dangerous mixture. In: *Iron Transport and Storage* (Ponka, P., Schulman, H. M., and Woodworth, R. C., Eds.), CRC Press, Boca Raton, FL, 1990.

Haber, R. and Weiss, J. The catalytic decomposition of hydrogen peroxide mixture by iron salts. *Proc. Soc. Lond. Br. Biol. Sci.* 147:332–351, 1934.

Halliwell, B. Antioxidants:sense or speculation? *Nutr. Today* 29:15–19, 1994.

Halliwell, B. and Gutteridge, J. M. C. *Free Radicals in Biology and Medicine,* Second Edition, Oxford Press, Oxford, 1989.

Halliwell, B., Gutteridge, J. M. C., and Cross, C. E. Free radicals, antioxidants, and human disease: Where are we now? *J. Lab. Clin. Med.* 119:598–620, 1991.

Hathcook, J. N. Safety and regulatory issues for phytochemical sources: "Designer Foods." *Nutr. Today* 28:23–33, 1993.

Heliovaara, M., Knekt, P., Aho, K. Aaran, R. K., Alfthan, G., and Aroma, A. Serum antioxidants and risk of rheumatoid arthritis. *Annals of Rheumatoid Dis.* 53:51–53, 1994.

Herbert, V., Shaw, S., and Jayatilleke, E. Vitamin C-driven free radical generation from iron. *J. Nutr.* 126:1213S–1220S, 1996.

Herbert, V., Shaw, S. Jayatilleke, E., and Stopler-Kasda, T. Most free radical injury is iron related: it is promoted by iron, hemin, holoferritin, and vitamin C, and inhibited by desferrioxamine and apoferritin. *Stem Cells* 12:289–303, 1994.

Ingold, K. U., Bowry, V. W., Stocker, R., and Walling, C. Autoxidation of lipids and antioxidation of α-tocopherol and ubiquinol in homogeneous solution and in aqueous dispersions of lipids: Unrecognized consequences of lipid particle size as examined by oxidation of human low density lipoprotein. *Proc. Natl. Acad. Sci.* 90:45–49, 1993.

Jacob, R. A. The integrated antioxidant system. *Nutr. Res.* 15:7555–7566, 1995.

Jacob, R. A., Skala, J. H., Omaye, S. T., and Hevia, P. Biochemical methods for assessing vitamin C status of the individual. In: *Nutritional Status Assessment of the Individual.* (Livingston, G. E., Ed.), Food and Nutrition Press, Inc, Trumbull, CT, pp. 323–337, 1989.

Jialal, I., Vega, G. L., and Grundy, S. M. Physiologic levels of ascorbate inhibit the oxidative modification of low density lipoprotein. *Atherosclerosis* 82:185–191, 1990.

Kagan, V. E., Serbinova, E. A., Forte, T., Scita, G., and Packer, L. Recycling of vitamin E in human low density lipoprotein. *J. Lipid Res.* 33:385–397, 1992.

Kamal-Eldin, A. and Appelqvist, L. A. The chemistry and antioxidant properties of tocopherols and tocotrienols. *Lipids* 31:671–703, 1996.

Kennedy, T. A. and Liebler, D. C. Peroxyl radical scavenging by β-carotene in lipid bilayers. Effect of oxygen partial pressure. *J. Biol. Chem.* 267:4658–4663, 1992.

Klaassen, C. D. *Casarett & Doull's Toxicology: The Basic Science of Poisons,* Fifth Edition, McGraw-Hill, New York, NY, pp. 13–34, 1996.

Kornbrust, D. J. and Mavis, R. D. Relative susceptibility of microsomes from lung, heart, kidney, brain, and testes to lipid peroxidation: Correlation with vitamin E content. *Lipids* 15:315–322, 1980.

Kostic, D., White, W. S., and Olson, J. A. Intestinal absorption, serum clearance, and interactions between lutein and β-carotene when administrated to human adults in separate or combined oral doses. *Am. J. Clin. Nutr.* 62:604–610, 1995.

Krinsky, N. I. Antioxidant functions of carotenoids. *Free Rad. Biol. Med.* 7:617–635, 1989.

Krinsky, N. I. Actions of carotenoids in biological systems. *Ann. Rev. Nutr.* 13:561–587, 1993.

Kubow, S. Routes of formation and toxic consequences of lipid peroxidation products in foods. *Free Rad. Biol. Med.* 12:63–81, 1992.

Kubow, S. Lipid oxidation products in food and atherogenesis. *Nutr. Rev.* 51:33–40, 1993.

Lauffer, R. B. Introduction, iron, aging, and human disease: historical background and new hypotheses. In: *Iron and Human Disease* (Lauffer, R. B., Ed.,) CRC Press, Boca Raton, FL, pp. 1–20, 1992.

Leibovitz, B., Hu, M. L., and Tappel, A. L. Dietary supplements of vitamin E, β-carotene, coenzyme Q_{10}, and selenium protect tissues against lipid peroxidation in rat tissue slices. *J. Nutr.* 120:97–104, 1990.

Liebler, D. C. Antioxidant reactions of carotenoids. *Ann. NY Acad. Sci.* 691:20–31, 1993.

Liebler, D. C. and McClure, T. D. Antioxidant reactions of β-carotene: Identification of carotenoid-radical adducts. *Chem. Res. Toxicol.* 9:8–11, 1996.

Lim, B. P., Nagao, A., Terao, J., Tanaka, K., and Suzuki, T. Antioxidant activity of xanthophylls on peroxyl radical mediated phospholipid peroxidation. *Biochem. Biophys. Acta.* 1126:178–184, 1992.

Lopez-Torres, M., Perez-Campo, R., Rojas, C., and Barja de Quiroga, C. Sensitivity to in vitro lipid peroxidation in liver and brain of aged rats. *Rev. Esp. Fisiol.* 48:191–196, 1992.

Machlin, L. J. *Vitamin E: A Comprehensive Treatise.* Marcel Dekker, Inc. New York, 1980.

Maritz, G. S. Ascorbic acid: Protection of lung tissue against damage. In: *Ascorbic Acid: Biochemistry and Biomedical Cell Biology, Subcellular Biochemistry Series.* (Harris, J. R., Ed.), Vol 25, Plenum Press, NY, pp. 265–292, 1996.

Mathews-Roth, M. M. β-Carotene therapy for erythropoietic protoporphyria and other photosensitivity diseases. *Biochimie* 68:875–884, 1986.

Mathews-Roth, M. M. Photoprotection by carotenoids. *Fed. Proc.* 46:1890–1893, 1987.

Mayne, S. T. and Parker, R. S. Subcellular distribution of β-carotene in chick liver. *Lipids* 21:164–169, 1986.

Mayne, S. T. and Parker, R. S. Antioxidant activity of dietary canthaxanthin. *Nutr. & Cancer* 12:225–236, 1989.

McLaran, C. J., Bett, J. H. N., Naye, J. A., and Halliday, J. W. Congestive cardiomy-opathy and hemochromatosis—rapid progression possibly accelerated by excessive ingestion of ascorbic acid. *Aust. N. Z. J. Med.* 12:187–188, 1982.

Meyskens, F. L. Coming of age—the chemoprevention of cancer. *N. Engl. J. Med.* 323:825–827, 1990.

Niki, E. Antioxidant compounds. In: *Chemical, Biological and Medical Aspects.* (Davis, K. J., Ed.), Pergamon Press, Elmsford, NY, pp. 57–64, 1991.

Niki, E., Noguchi, N., Tsuchihashi, H., and Gotoh, N. Interactions among vitamin C, vitamin E and β-carotene. *Am. J. Clin. Nutr.* 62:1322S–6S, 1995.

O'Connell, M. J., Halliwell, B., Moorhouse, C. P., Aryoma, D. I., Baum, H., and Peters, T. J. Formation of hydroxyl radicals in the presence of ferritin and hemosiderin. *Biochem J.* 234:727–733, 1986.

Olcott, H. S. and Van Der Ven, J. Comparison of antioxidant activities of tocol and its methyl derivatives. *Lipids* 3:331–334, 1968.

Olson, J. A. Biological actions of carotenoids. *J. Nutr.* 119:94–95, 1989.

Olson, J. A. Benefits and liabilities of vitamin A and carotenoids. *J. Nutr.* 126:1208S–1212S, 1996.

Olson, J. A. Vitamin A. In: *Present Knowledge in Nutrition.* (Ziegler, E. E. and Filer, L. J., Jr, eds.), ILSI Press, Washington, D.C., pp. 109–119, 1996.

Olson, J. A. and Krinsky, N. I. The colorful fascinating world of the carotenoids: Important physiologic modulators. *FASEB J.* 9:1547–1550, 1995.

Omaye, S. T., Elsayed, N. M., Klain, G. J., and Korte, D. W., Jr. Metabolic changes in the mouse kidney after a subcutaneous injection of butyl 2-chloroethyl sufide. *J. Toxicol. Environ. Health* 33:19–27, 1991.

Omaye, S. T., Tillotson, J. A., and Sauberlich, H. E. Metabolism of L-ascorbic acid in the monkey. In: *Adv. Chem. Series,* (Seib, P. A. and Tolbert, B. M., Eds.), ACS, Washington, D.C., pp 317–334, 1982.

Omaye, S. T., Turnbull, J. D., and Sauberlich, H. E. Selected methods for the determination of ascorbic acid in animal cells, tissues, and fluids. *Meths. Enzymol.* 62:3–11, 1979.

Omenn, G. S., Goodman, G. E., Thornquist, M. D., Balmes, J., Cullen, M. R., Glass, A., Keogh, J. P., Meyskens, F. L., Jr., Valanis, B., Williams, J. H., Jr., Barnhart, S., and Hammar, S. Effects of combination of β-carotene and vitamin A on lung cancers in male smokers. *N. Engl. J. Med.* 334:1150–1155, 1996.

Ozhogina, O. A. and Kasaikina, O. β-Carotene as an interceptor of free radicals. *Free Rad. Biol. Med.*, 19:575–581, 1995.

Packer, L. and Fuchs, J. *Vitamin E in Health and Disease,* Marcel Dekker, New York, 1993.

Palozza, P., Calviello, G., and Bartoli, G. M. Prooxidant activity of β-carotene under 100% oxygen pressure in rat liver microsomes. *Free Rad. Biol. Med.* 19:887–892, 1995.

Palozza, P. and Krinskry, N. I. . The inhibition of radical initiated peroxidation of microsomal lipids by both α-tocopherol and β-carotene. *Free Rad. Biol. Med.* 11:407–414, 1991.

Palozza, P. and Krinsky, N. I. β-carotene and α-tocopherol are synergistic antioxidants. *Arch. Biochem. Biophys.* 297:184–187, 1992.

Parker, R. S. Absorption, metabolism, and transport of carotenoids. *FASEB J.* 10:542–551, 1996.

Peers, K. E. and Coxon, D. T. Controlled synthesis of monohydroperoxides by α-tocopherol inhibited autoxidation of polyunsaturated lipids. *Chem. Phys. Lipids* 32:49–56, 1983.

Pellett, L. J., Andersen, H. J., Chen, H., and Tappel, A. L. β-Carotene alters vitamin E protection against heme protein oxidation and lipid peroxidation in chicken liver slices. *J. Nutr. Biochem.* 5:479–484, 1994.

Pokorny, J. Major factors affecting the autoxidation of lipids. In: *Autoxidation of Unsaturated Lipids* (Chan, H. W. S., ed.), pp. 141–206, Academic Press, London, 1987.

Poovaiah, B. P. and Omaye, S. T. Response of plasma retinol to dietary changes of ascorbic acid and selenium in the guinea pig. *Nutr. Res.* 6:583–588, 1986.

RDA. *Recommended Dietary Allowances,* 10th Edition, National Research Council, National Academy Press, Washington, D.C., 1989.

Reardon, E. M. Phytochemicals and health. *Plant Mol. Biol. Rep.* 13:293–297, 1995.

Reaven, P. D., Ferguson, E., Navab, M., and Powell, F. L. Susceptibility of human LDL to oxidative modification. *Arterioscler. Thromb.* 14:1162–1169, 1994.

Reddy, A. K. and Omaye, S. T. Target organ toxicity and metabolic and biochemical responses following lung exposure. In: *Pulmonary Toxicology.* (Salem, H., Ed.), Marcel Dekker, New York, NY, pp. 223–253, 1987.

Rose, R. C. and Bode, A. M. Biology of free radical scavengers: An evaluation of ascorbate. *FASEB J.* 7:1135–1142, 1993.

Ryan, T. P. and Aust, S. D. The role of iron in oxygen-mediated toxicities. *Crit. Rev. Toxicol.* 22:119–141, 1993.

Salonen, J. T., Salonen, R., Nyyssomen, K., and Korpela, H. Iron sufficiency is associated with hypertension and excess risk of myocardial infarction: The Kupio Ishemic Heart Disease Risk Factor Study (KIHD). *Circulation* 85:864–876, 1992.

Samokyszyn, V. M. and Marnett, L. J. Hydroperoxide-dependent cooxidation of 13-*cis*-retinoic acid by prostaglandin H synthetase. *J. Biol. Chem.* 262: 14119–14133, 1987.

Sauberlich, H. E., Green, M. D., and Omaye, S. T. Determination of ascorbic acid and dehydroascorbic acid. *Advances in Chemistry Series,* (Seib, P. A. and Tolbert, Eds.), ACS, Washington, D.C., pp. 199–221, 1982.

Sevanian, A., Hacker, A. D., and Elsayed, N. Influence of vitamin E and nitrogen dioxide on lipid peroxidation in rat lung and liver microsomes. *Lipids* 17:269–277, 1982.

Sies, H. Oxidative stress: Introductory remarks. In: *Oxidative Stress,* (Sies, H., ed.), New York: Academic Press, pp. 1–10, 1985.

Sies, H. and Stahl, W. Vitamin E and C, β-carotene, and other carotenoids as antioxidants. *Am. J. Clin. Nutr.* 62:1315S–21S, 1995.

Sokol, R. J. Vitamin E. In: *Present Knowledge in Nutrition.* (Ziegler, E. E. and Filer, L. J., Jr., eds)., ILSI Press, Washington, D.C., pp. 130–136, 1996.

Stadman, E. R. Ascorbic acid and oxidative inactivation of proteins. *Am. J. Clin. Nutr.* 54:1125S–1128S, 1991.

Stratton, S. P., Schaefer, W. H., and Liebler, D. C. Isolation and identification of singlet oxygen oxidation products of β-carotene. *Chem. Res. Toxicol.* 6:542–547, 1993.

Takahashi, M., Tsuchiya, J., and Niki, E. Scavenging of radicals by vitamin E in the membranes as studied by spin labeling. *J. Am. Chem. Soc.* 111:6350–3, 1989.

Takahashi, M., Tsuchiya, J., Niki, E., and Urano, S. Action of vitamin E as antioxidant in phospholipid liposomal membranes as studied by spin label technique. *J. Nutr. Sci. Vitaminol.* 34:24–34, 1988.

Tappel, A. L. Vitamin E as the biological lipid antioxidant. *Vitamins and Hormones: Advances in Research and Applications,* 20:493–510, 1962.

Tappel, A. L. Vitamin E. *Nutr. Today* 8:4-12, 1973.

Tappel, A. L. Combinations of vitamin E and other antioxygenic nutrients in protection of tissues. In: *Vitamin E in Health and Disease.* (Packer, L. and Fuchs, J., Eds.), Marcel Dekker, New York, NY, pp. 313–325, 1993.

Terao, J. and Matsushita, S. The peroxidizing effect of α-tocopherol on autoxidation of methyl linoleate in bulk phase. *Lipids* 21:255–260, 1986.

α-Tocopherol, β-Carotene Cancer Prevention Study Group. The effects of vitamin E and β-carotene on the incidence of lung cancer and other cancers in male smokers. *N. Engl. J. Med.* 330:1029–1035, 1994.

Toyokuni, S. and Sagripanti, J. L. Iron-mediated DNA damage: Sensitive detection of DNA strand breakage-catalyzed by iron. *J. Inorg. Biochem.* 47:241–248, 1992.

Von Zglinicki, T., Wiswedel, I., Trumper, L., and Agustin, W. Morphological changes of isolated rat liver mitochondria during Fe^{2+}/ascorbate-induced peroxidation and the effect of thioctic acid. *Mech. Aging Dev.* 57:233–246, 1991.

Wang, X.-D. Absorption and metabolism of β-carotene. *J. Am. Col. Nutr.* 13: 314–325, 1994.

Wardlaw, G. M., and Insel, P. M. *Perspectives in Nutrition,* Third Edition, Mosby, St. Louis, MO, pp. 410–411, 1996.

White, W. S., Peck, K. M., Bierer, T. L., Gugger, E. T., and Erdman, J. W., Jr. Interactions of oral β-carotene and canthaxanthin in ferrets. *J. Nutr.* 123:1405–1513, 1993.

Will, O. H., III and Scovel, C. A. Photoprotective functions of carotenoids. In: *Carotenoids: Chemistry and Biology,.* (Krinsky, N. I., Mathews-Roth, M. M., and Taylor, R. F., eds.), Plenum Press, New York, pp. 229–36, 1989.

Willett, W. C., Stampfer, M. J., Undersood, B. A., Taylor, J. O., and Hennekins, C. H. Vitamin A, E, and carotene: Effects of supplementation on their plasma levels. *Am. J. Clin. Nutr.* 38:559–566, 1983.

Wilson, R. L. Free radical protection: Why vitamin E, not vitamin C, beta-carotene or glutathione? *Cib. Found. Symp.* 101:19–44, 1983.

Wiseman, H. Dietary influences on membrane function: Importance in protection against oxidative damage and disease. *Nutr. Biochem.* 7:2–15, 1996.

Wolf, G. Is β-carotene an anticancer agent? *Nutr. Rev.* 40:257–261, 1982.

Woodall, A. A., Britton, G., and Jackson, M. J. Dietary supplementation with carotenoids: Effects on α-tocopherol levels and susceptibility of tissues on oxidative stress. *Br. J. Nutr.* 76:307–317, 1996.

Yip, R. Changes in iron metabolism with age. In: *Iron Metabolism in Health and Disease,* (Brock, J. H., Halliday, J. W., Pippard, M. J., and Powell, L. W., Eds.), Philadelphia: W. B. Saunders Company Ltd. pp. 427–448, 1994.

CHAPTER 5

Distribution, Bioavailability, and Metabolism of Carotenoids in Humans

FREDERICK KHACHIK
FREDERICK B. ASKIN
KEITH LAI

INTRODUCTION

To date, approximately 600 carotenoids have been isolated from natural sources, and their structures have been characterized.[1] However the number of dietary carotenoids in common fruits and vegetables consumed in the United States is in excess of 40.[2,3] Only 13 all-*trans*-carotenoids belonging to the class of hydroxy- and hydrocarbon carotenoids (carotenes) have been detected in human plasma. In addition, the structures of eight metabolites resulting from 2 major dietary carotenoids, namely, lutein and lycopene, have been elucidated.[4-8] Among plasma carotenoids, only α-carotene, β-carotene, β-cryptoxanthin, and γ-carotene can be converted to vitamin A, although the nutritional significance of other non-vitamin A-active carotenoids in prevention of cancer,[9,10] heart disease,[11,12] and age-related macular degeneration[13] (AMD) has been suggested.

Here we describe a summary of our most recent findings with regard to the distribution of carotenoids in human plasma, breast milk, as well as several organs and tissues including liver, lung, breast, cervix, and retina. The bioavailability and metabolism of carotenoids in humans supplemented with purified carotenoids and subjects consuming a diet rich in fruits and vegetables will be presented. Based on these human bioavailability studies, the antioxidant mechanism of action for specific carotenoids, such as lutein, zeaxanthin, and lycopene, as potential candidates to lower the risk for cancers and AMD is discussed.

EVIDENCE FOR NUTRITIONAL BENEFIT OF CAROTENOIDS IN DISEASE PREVENTION

The evidence for nutritional prevention of cancers, heart disease, and AMD by carotenoids has been obtained from various interdisciplinary studies; these may be classified as (a) epidemiological studies, (b) carotenoid distribution in fruits, vegetables, human serum, and breast milk, (c) carotenoids in human organs and tissues, (d) *in vitro* studies of chemopreventive properties, and (e) a recent *in vivo* study with rodents. Each of these studies will be briefly discussed.

EPIDEMIOLOGICAL STUDIES

Numerous epidemiological studies have demonstrated that consumption of carotenoid-rich fruits and vegetables lowers the risk for certain cancers in various human populations. For a comprehensive review, see the article by van Poppel.[10] With regard to heart disease, there are at least two recent studies that have indicated the protective role of carotenoids.[11,12] A recent eye disease case-controlled study of AMD investigated the effect of consuming specific carotenoids on the prevalence of this disease.[13] A total of 520 controls and 356 cases with AMD, ages 55 to 80 years, participated. The results indicated that the prevalence of AMD was reduced by 43% in the quintile consuming 6 mg/day of lutein from fruits and vegetables. The authors concluded that the consumption of lutein-rich foods may help to reduce the risk for AMD. Our recent results on distribution of carotenoids in human retina also suggest a protective role of lutein and zeaxanthin in preventing AMD; these will be discussed later.

CAROTENOID DISTRIBUTION IN FRUITS, VEGETABLES, HUMAN SERUM, AND BREAST MILK

In addition to their widespread distribution in fruits and vegetables, carotenoids are also present in human serum and breast milk.[14] The carotenoids of fruits and vegetables can be classified as (1) hydrocarbon carotenoids or carotenes, (2) monohydroxycarotenoids, (3) dihydroxycarotenoids, (4) carotenol acyl esters, and (5) carotenoid epoxides. For a comprehensive review of the distribution of carotenoids in fruits and vegetables, see the articles by Khachik et al.[14–16] The concentrations of carotenoids in fruits and vegetables commonly consumed in the United States have also been reported.[16–20] The effect of mild cooking and processing on qualitative and quantitative distribution of carotenoids in selected fruits and vegetables has revealed no significant changes in carotenoid profiles.[19,20]

A complete list of dietary carotenoids and their metabolites that have been detected in human serum and breast milk is provided in Table 1.[14] These

TABLE 1. Dietary Carotenoids and Their Metabolites in Human Serum
and Breast Milk.

Entry	Carotenoids	Chemical Class
all-trans-Dietary Carotenoids*		
1	α-carotene*	Hydrocarbons (carotenes)
2	β-carotene*	
3	γ-carotene*	
4	lycopene	
5	neurosporene	
6	ζ-carotene	
7	phytofluene	
8	phytoene	
9	α-cryptoxanthin	Monohydroxycarotenoids
10	β-cryptoxanthin*	
11	lutein	Dihydroxycarotenoids
12	zeaxanthin	
13	lactucaxanthin	
cis-Dietary Carotenoids*		
14	9-*cis*-β-carotene*	Hydrocarbons (carotenes)
15	13-*cis*-β-carotene*	
16	*cis*-lycopenes	
17	*cis*-phytofluenes	
18	*cis*-β-cryptoxanthin*	Monohydroxycarotenoid
19	13,13'-*cis*-lutein	Dihydroxycarotenoids
20	9-*cis*-lutein	
21	9'-*cis*-lutein	
22	13-*cis*-lutein + 13'-*cis*-lutein	
23	9-*cis*-zeaxanthin	
24	13-*cis*-zeaxanthin	
25	15-*cis*-zeaxanthin	
cis- and *all-trans*-Carotenoid Metabolites		
26	3'-hydroxy-ϵ,ϵ-caroten-3-one	Monoketocarotenoid
27	3-hydroxy-β,ϵ-caroten-3'-one	
28	*cis*-3-hydroxy-β,ϵ-caroten-3'-one	
29	ϵ,ϵ-caroten-3,3'-dione	Diketocarotenoid
30	3-hydroxy-3',4'-didehydro-β,γ-carotene	Monohydroxycarotenoids
31	3-hydroxy-2',3'-didehydro-β,ϵ-carotene	
32	3'-epilutein	Dihydroxycarotenoids
33	2,6-cyclolycopene-1,5-diol A	
34	2,6-cyclolycopene-1,5-diol B	

*Identifies carotenoids with vitamin A activity.

79

include 34 carotenoids consisting of 25 dietary (13 all-*trans*- and 12 *cis*-compounds) and 9 carotenoid metabolites (1 *cis*- and 8 all-*trans*-compounds). Their chemical structures are shown in Figure 1.

Comparison of the qualitative profile of carotenoids in foods with those of human serum and breast milk is shown in Figure 2. The first group are vitamin A-active carotenoids such as α-carotene, β-carotene, β-cryptoxanthin, and γ-carotene.

The second group, consisting of α-cryptoxanthin, neurosporene, ζ-carotene, phytofluene, and phytoene, has no vitamin A activity and appears to be absorbed intact. At present, there is no evidence to suggest that these carotenoids undergo metabolic transformation.

Several oxidative metabolites of lycopene, lutein, zeaxanthin, and lactucaxanthin in human serum and breast milk have been isolated and characterized (compounds 26–29 and 32–34, Figure 1).[14] Human bioavailability and metabolic studies have supported the possibility of *in vivo* oxidation of these carotenoids in humans.[8,21–22] The metabolic transformation of these carotenoids involves a series of oxidation-reduction reactions, which have been described previously.[8,21] There are also two metabolites of lutein that are apparently formed in the presence of acids by nonenzymatic dehydration of this compound in the human digestive system (compounds 30 and 31, Figure 1).[7]

Carotenol acyl esters are abundant in many fruits and vegetables but have not been detected in human serum.[2,17–18] Carotenol acyl esters are believed to undergo hydrolysis in the presence of pancreatic secretions to regenerate their parent hydroxycarotenoids, which are then absorbed. These are mono- and dihydroxycarotenoids, which are esterified with lauric, myristic, and palmitic acids in certain foods. Only two monohydroxy carotenoids, α-cryptoxanthin and β-cryptoxanthin, have been detected in common fruits and vegetables as well as human plasma. Similarly, of all the dihydroxycarotenoids isolated from various natural sources, only lutein, zeaxanthin, and lactucaxanthin have been found in foods and human serum.[4,14] The concentration of lactucaxanthin in human plasma is normally very low because the dietary source of this compound is limited, and to date we have detected it only in romaine lettuce (*Lactuca sativa*).[14]

The most common carotenoid epoxides, widely present in foods, are neoxanthin, violaxanthin, lutein 5,6-epoxide, and β-cryptoxanthin epoxide. However, to date, none of these compounds has been detected in human serum or plasma.[2]

CAROTENOIDS IN HUMAN ORGANS AND TISSUES

Preliminary analysis of extracts from several human organs and tissues including liver, lung, breast, and cervix by high-performance liquid chromatography (HPLC) has revealed the presence of several major dietary carotenoids.

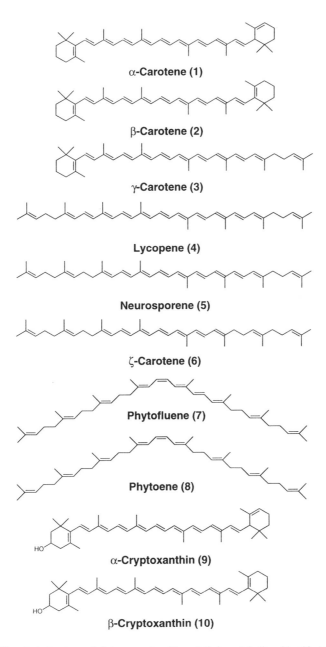

α-Carotene (1)

β-Carotene (2)

γ-Carotene (3)

Lycopene (4)

Neurosporene (5)

ζ-Carotene (6)

Phytofluene (7)

Phytoene (8)

α-Cryptoxanthin (9)

β-Cryptoxanthin (10)

Figure 1 Chemical structure of dietary carotenoids and their metabolites identified in human serum, milk, organs, and tissues. With the exception of phytofluene and phytoene, only the all-*trans* isomers are shown. Only the absolute configuration rather than the planar structure for those carotenoids with confirmed chirality have been depicted. An asterisk (*) indicates that only the relative and not absolute configuration at C-2, C-5, and C-6 for compounds 33 and 34 are known. Where possible, the common names for certain carotenoids have been used. Numbers associated with the carotenoids correspond to the complete list of carotenoids shown in Table 1. (Reprinted permission from Reference 14. Copyright 1997 American Chemical Society.)

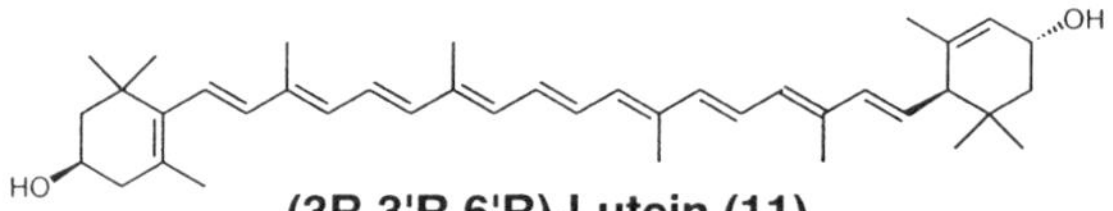

(3R,3'R,6'R)-Lutein (11)

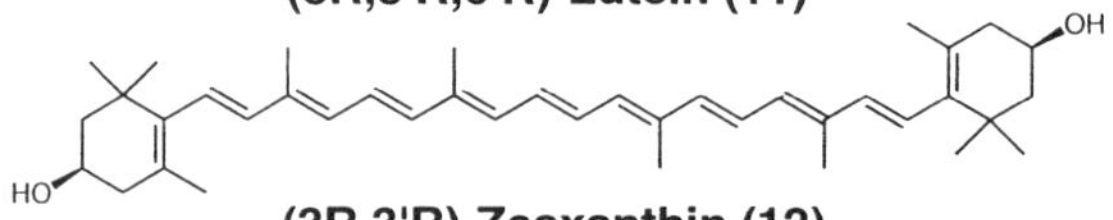

(3R,3'R)-Zeaxanthin (12)

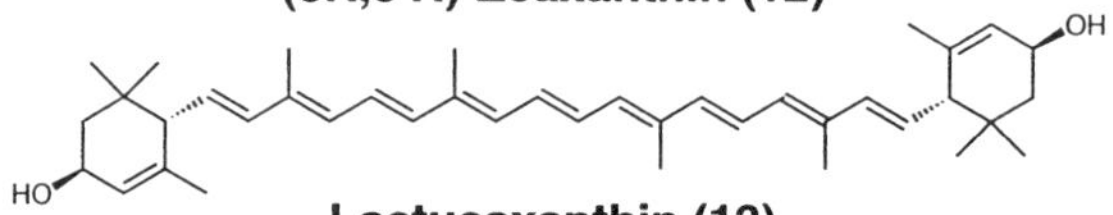

Lactucaxanthin (13)

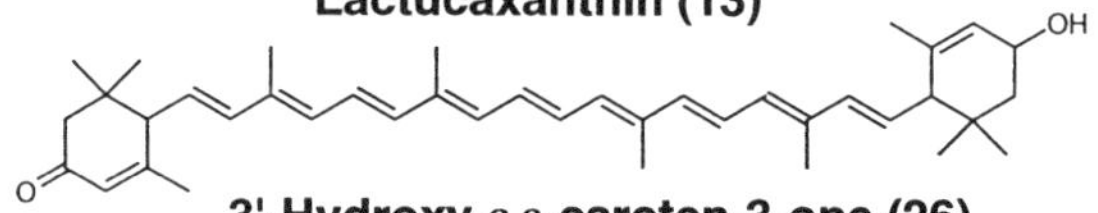

3'-Hydroxy-ε,ε-caroten-3-one (26)

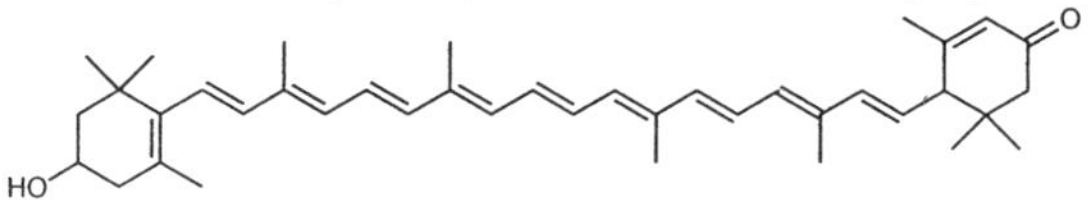

3-Hydroxy-β,ε-caroten-3'-one (27)

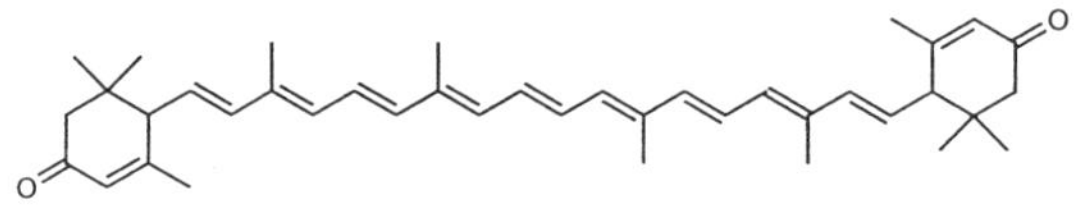

ε,ε-Caroten-3,3'-dione (29)

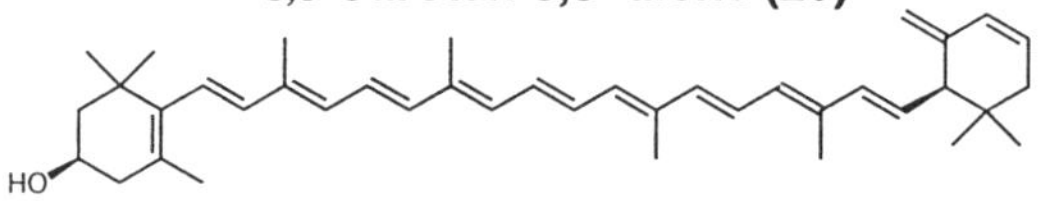

3-Hydroxy-3',4'-didehydro-β,γ-carotene (30)

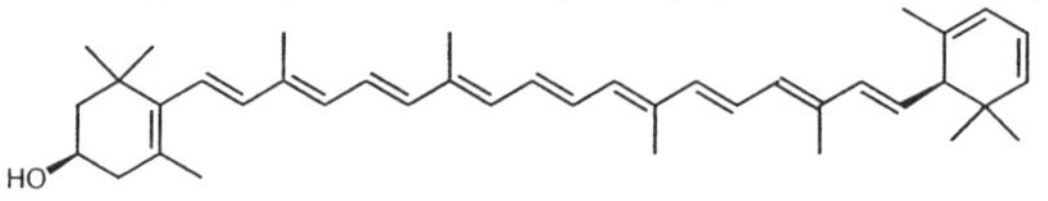

3-Hydroxy-2',3'-didehydro-β,ε-carotene (31)

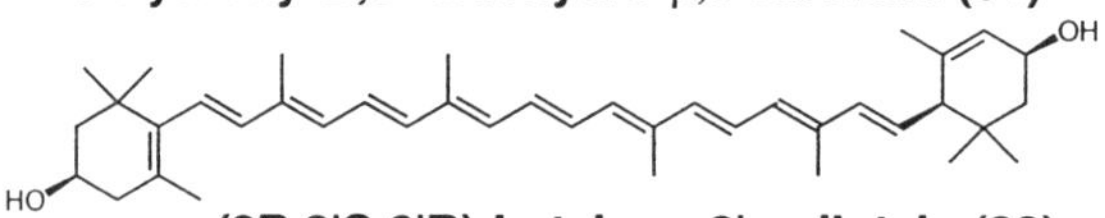

(3R,3'S,6'R)-Lutein or 3'-epilutein (32)

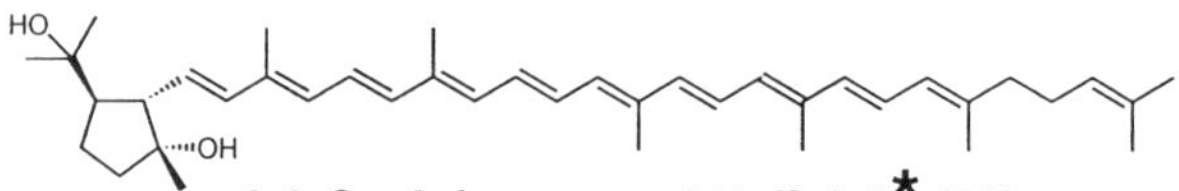

2,6-Cyclolycopene-1,5-diol A* (33)

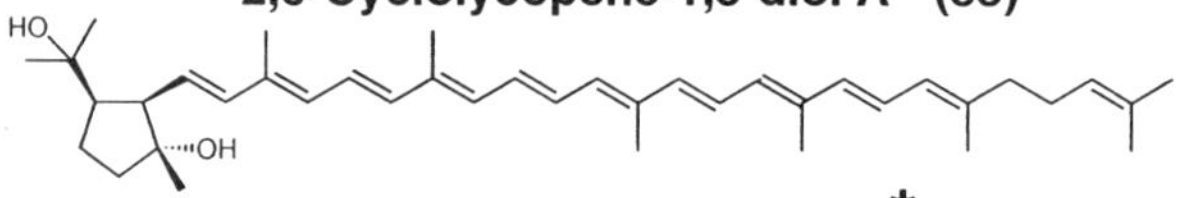

2,6-Cyclolycopene-1,5-diol B* (34)

Figure 1 *(continued).*

82

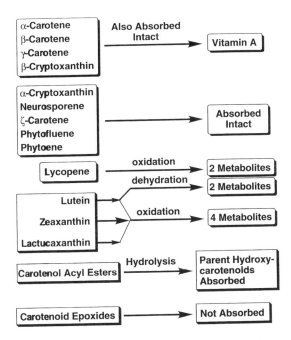

Figure 2 Comparison between dietary and serum carotenoids and their metabolic transformation in humans.

The specimens were collected at the time of surgery by a senior pathologist and the tumor bank technician at Johns Hopkins Medical Institution (Baltimore, MD). In each case, only grossly normal portions of the specimen were collected, and all tissue collected was taken from areas that were determined not to be necessary for diagnosis or patient care decisions. All tissues were provided to the Beltsville Human Nutrition Research Center (BHNRC) without identifiers that would allow discovery of the patient's name. All samples were carefully separated from the main specimen by sharp dissection and were frozen immediately in liquid nitrogen. They were stored at −20°C and transported to BHNRC on dry ice at approximately −40°C. The history of the patient in each case is briefly described as follows:

(1) Liver: Sample was obtained from a 65-year-old female with primary diagnosis of metastatic adenocarcinoma in the liver (primary site = gallbladder), and the surgical procedure was segmental resection of liver.

(2) Lung: Sample obtained from a 55-year-old female with primary diagnosis of lung carcinoma, and the surgical procedure was lobectomy.

(3) Breast: Sample obtained from a 72-year-old female with primary diagnosis of breast cancer. The surgical procedure was modified radical mastectomy.

(4) Cervix: Sample was obtained from a 47-year-old female with primary diagnosis of uterine leiomyomas. The surgical procedure was total abdominal hysterectomy and bilateral salpingo-oophorectomy.

Extraction and HPLC-MS Analysis

Samples were thawed and extracted by homogenization with tetrahydrofuran in the presence of methanolic KOH (10%) for 2 h followed by partitioning into hexane and water. The hexane layer was removed, dried over sodium sulfate, and evaporated to dryness. The extracts were reconstituted in the appropriate solvents, and carotenoids were separated on a C_{18}-reverse-phase HPLC column by photodiode array detection (UV-VIS)-particle beam mass spectrometry (MS) similar to that of our published procedure.[4] Identification of carotenoids was accomplished by comparison of their HPLC-UV-VIS-MS profile with those of synthetic and purified carotenoid standards.

Results

The concentrations of carotenoids in human liver, lung, breast, and cervix are shown in Table 2. It appears that all the dietary carotenoids that are present in serum are also accumulated in these organs and tissues. The extracts of these specimens were not investigated for the presence of oxidative metabolites of carotenoids because alkaline hydrolysis during extraction usually results in the degradation of these compounds. This problem can be eliminated by employing enzymatic hydrolysis.[14] Although accumulation of β-carotene in human liver has been well documented, to our knowledge, this is the first report that provides evidence for the hepatic storage of other dietary and serum carotenoids. 2,6-Cyclolycopene-1,5-diol is a metabolite of lycopene, which is also found at very low concentrations in tomato products. The presence of this compound in human liver, serum, and milk might be a result of the *in vivo* oxidation of lycopene.[8] With regard to carotenoid distribution in lung tissue, the relatively high concentration of lutein, β-cryptoxanthin, lycopene, β-carotene, phytofluene, and particularly phytoene is notable. It is imperative to point out that the major dietary sources of lycopene, ζ-carotene, phytofluene, and phytoene are tomatoes and tomato products.[19,20] Over the years, our extensive serum analysis of human subjects with various dietary regimens has revealed that the relative concentration of serum carotenoids is reflective of their ratio in tomatoes and tomato products. Therefore, the unusually high concentration of phytoene (Table 2), relative to the other tomato carotenoids, may be due to a preferential uptake of this com-

TABLE 2. Carotenoids Identified in Human Liver, Lung, Breast, and Cervix.*

Carotenoids	Concentration (ng/g)			
	Liver	Lung	Breast	Cervix
Lutein	343.5	298.2	130.3	23.8
Anhydrolutein	64.4	—b	—b	—b
α-Cryptoxanthin	126.7	30.9	23.1	4.0
β-Cryptoxanthin	363.1	120.7	37.1	24.3
Lycopene	351.8	299.5	233.5	95.0
2,6-Cyclolycopene-1,5-diol	21.1	—**	—**	—**
ζ-Carotene	150.1	25.1	734.4	57.2
α-Carotene	67.2	47.4	128.2	23.6
β-Carotene + *cis*-isomers	470.0	225.6	356.3	125.3
Phytofluene	261.4	194.6	416.2	106.3
Phytoene	167.8	1,274.5	68.9	—**
Total	2,387.1	2,516.5	2,128.0	459.5

*All organs and tissues contained several unidentified breakdown products of carotenoids which were not characterized.
**Not detectable.

pound by the lung tissues. Similarly, high concentrations of ζ-carotene and phytofluene in breast tissue relative to the other carotenoids are particularly interesting. Major carotenoids in cervical tissue appear to be lycopene, β-carotene, and phytofluene; surprisingly, no phytoene could be detected. In all cases, the analysis of the extracts from various organs and tissues also revealed the presence of several unidentified apparent degradation products of carotenoids. The data shown in Table 2 must be considered tentative because these results have been obtained from multiple subjects with a single sample in each case. Analysis of large numbers of samples will be needed to establish the normal tissue concentration of specific carotenoids and their metabolites.

Lutein, Zeaxanthin, and Their Oxidative Metabolites in Human Retina

In 1985, for the first time, Bone et al.[23] presented preliminary evidence that the human macular pigment is a combination of lutein and zeaxanthin. Employing HPLC, Bone et al. further studied the retinal distribution of lutein and zeaxanthin for 87 donors between the ages of 3 and 95.[24] The lutein-zeaxanthin ratio increased in individual retinas from an approximate average of 1:2.4 in the central (0–0.25 mm) to more than 2:1 in the periphery (8.7–12.2 mm).[24] More recently, in 1993, Bone et al.[25] elegantly established the complete identification and stereochemistry of the human macular pigment as lutein [(3R,3′R,6′R)-β,ϵ-carotene-3,3′diol], zeaxanthin [(3R,3′R)-β,β-carotene-3-3′diol], and *meso*-zeaxanthin [(3R,3′S)-β,β-carotene-3,3′diol]. In a case-controlled study published in 1994, high consumption of fruits and vegetables, rich specifically in lutein and zeaxanthin, has been correlated with a lower risk for AMD.[13] The results from that study indicated that persons consuming a diet containing a high concentration of lutein (approximately 6 mg/day) from leafy green vegetables had 43% lower risk of exudative AMD compared to the subjects in the lowest quintile.

Recently, we have separated and characterized 3 major and 11 minor carotenoids in human retina by HPLC-UV-vis-MS.[26] The major carotenoids have been identified as lutein, zeaxanthin, and a direct oxidation product of lutein, 3-hydroxy-β,ϵ-caroten-3′-one (compound 27, Figure 1). Several oxidation products of lutein and zeaxanthin (compounds 26, 29, and 32, Figure 1) and one of lycopene (compound 33, Figure 1) were also among the minor carotenoids. In addition, the most common geometrical isomers of lutein and zeaxanthin, i.e., 9′-*cis*-lutein, 9′-*cis*-lutein, 13-*cis*-lutein, 13′-*cis*-lutein, 9-*cis*-zeaxanthin, and 13-*cis*-zeaxanthin, normally present in serum, were also detected at low concentrations in retina. Similar results were also obtained from an extract of freshly dissected retina of a healthy monkey. At present, this finding provides the only primate data, suggesting *in vivo* metabolic oxidation of lutein in human retina. The protection of retina from short-wavelength visible light by lutein and zeaxanthin is based on two assumptions: (1) the oxidation products of lutein and zeaxanthin are formed *in vivo* in retina and (2) these oxidative metabolites are formed by the action of blue light. Based on supplementation studies with purified lutein and zeaxanthin involving human subjects, we previously proposed metabolic pathways for conversion of these dietary carotenoids to their oxidation products.[8,21] According to these metabolic transformations, (3R,3′S,*meso*)-zeaxanthin, (3R,3′R,6′R)-lutein, and (3R,3′R)-zeaxanthin may be interconverted through a series of oxidation-reduction and double bond isomerization reactions, as shown in Figure 3. The driving force for the direct oxidation of lutein to 3-hydroxy-β,ϵ-caroten-3′-one is the activation of the hydroxyl group at C-3′ by the neighboring allylic double bond. However in (3R,3′R)-zeaxanthin, due to the nonallylic nature of the hydroxyl groups at C-3 and C-3′, this compound is not directly oxidized and may therefore undergo double-bond isomerization to yield 3′-epilutein [(3R,3′S,6′R)-lutein] prior to oxidation. As a result, 3-hydroxy-β,ϵ-caroten-3′-one can also be formed from allylic oxidation of 3′-epilutein. To establish interconversion between carotenoids in these metabolic transformations, once 3-hydroxy-β,ϵ-caroten-3′-one is formed, it may undergo reduction reactions with epimerization at C-3′ to yield lutein and/or 3′-epilutein. In another double-bond isomerization reaction, dietary (3R,3′R,6′R)-lutein may be transformed into (3R,3′R,*meso*)-zeaxanthin (Figure 3). We have recently deter-

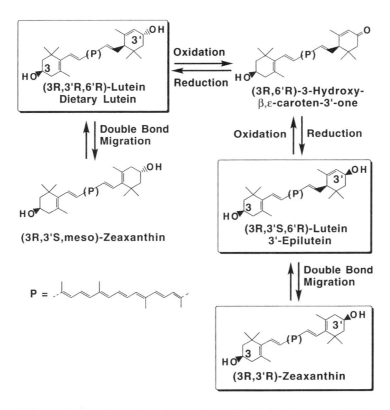

Figure 3 Proposed metabolic transformations of dietary (3R,3'R,6'R)-lutein and (3R,3'R)-zeaxanthin in human retina. (Reprinted with permission from reference 26. Copyright 1997 Association for Research in Vision and Ophthalmology.)

mined that 98% of zeaxanthin in human plasma is identical to the form in the diet, (3R,3'R)-zeaxanthin; only 2% exists as (3R,3'R,*meso*)-zeaxanthin (Khachik, unpublished results). Because Bone et al.[25] have demonstrated that nearly equal amounts of (3R,3'R)-zeaxanthin and (3R,3'R,*meso*)-zeaxanthin are present in the human macula, the data indicate that the latter compound may be a result of double-bond isomerization of dietary (3R,3'R,6'R)-lutein, as shown in Figure 3. However, the only data in support of these metabolic transformations is the presence of 3-hydroxy-β,ϵ-caroten-3'-one, 3'-epilutein, and *meso*-zeaxanthin in retina. The transport and the metabolic interconversions between lutein and the two forms of zeaxanthin in human retina could be induced by the radiation from short-wavelength visible light and/or catalyzed by certain proteins. Human macular carotenoids bind to retinal tubulin,[27] and other more specific human macular carotenoid binding proteins are being identified and characterized (Bernstein et al., unpublished results).

IN VITRO STUDIES OF CHEMOPREVENTIVE PROPERTIES

Carotenoids also exhibit some of the biological activities of chemopreventive agents. Chemoprevention may be defined as the inhibition or reversal of carcinogenesis, a process that begins with cells of normal morphology and ends with invasive cancer. Compounds that inhibit carcinogenesis become chemopreventive agents. The concept of chemoprevention of certain diseases such as cancer using phytochemicals is promising because the risk for many cancers may be lowered by optimum nutrition. Furthermore, most of the phytochemicals originating from foods have no toxicity or side effects. The duration between onset of cancer or preinitiation stage to the formation of invasive or clinical cancer may take 10 to 15 years or longer; as a result, this provides a sufficient opportunity for intervention with appropriate chemopreventive agents. Possible mechanisms by which chemopreventive agents work against events that are involved in development of cancer have been reviewed by Kelloff et al.[28] These events include formation and activation of carcinogens, induction of genetic damage, stimulation of cell proliferation, and disruption of normal cell growth and differentiation regulation.[29] For example, carotenoids, due to their antioxidant properties, may prevent the genetic damage by scavenging reactive electrophiles and protect the DNA from oxidative damage. This would be expected to result in the formation of a number of oxidative metabolites of carotenoids that have been detected in the human serum or plasma.[4,8,14,21] We have recently shown *in vitro* that carotenoids, and particularly their metabolites, are inducers of phase II metabolizing or detoxification enzymes (unpublished results). Enhancement of detoxification enzymes by chemopreventive agents increases intracellular glutathione levels, which is known to trap reactive electrophiles and free radicals and thereby prevent these species from entering into damaging interactions with the DNA.[30]

Perhaps the most well-established role of carotenoids as chemopreventive agents is their ability to restore tumor suppressor function and/or inhibit oncogene expression by increasing cellular communication. Using an *in vitro* murine model of carcinogenesis, Bertram has shown that carotenoids inhibit the postinitiation phase of both physically (x-ray) and chemically (methylcholanthrene) induced carcinogenesis.[31] This activity has been correlated with up-regulated gap junctional communication driven by up-regulated expression of connexin43 at the message and protein level. Based on these results, it has been proposed that increased junctional communication between growth-inhibited normal cells and carcinogen-initiated cells acts to place these initiated cells in a growth-inhibited environment and suppresses their ability to undergo neoplastic transformation. More recently, King et al.[32] have shown that 3 major dietary carotenoids, β-carotene, lutein, and lycopene, can increase connexin43 gene expression in 10T1/2 cells and in human keratinocytes in organotypic culture.[32] In addition, (3R,3'R)-zeaxanthin, a dietary carotenoid, and 2,6-cyclolycopene-1,5-diol, an oxidative metabolite of lycopene, were shown to exhibit the greatest activity in comparison with other carotenoids and their metabolites.

A possible mechanism underlying the positive association between dietary intake of carotenoid-rich fruits and vegetables and decreased incidence of certain cancers may be related to enhanced immune responses by carotenoids,[33,34] noncarotenoid phytonutrients, or both.[35,36] Although studies in animals show an immunoenhancing effect by isolated β-carotene,[37,38] similar findings in humans have been inconsistent.[39–41] After comparing the different cell-culture methods used to measure the effects of carotenes on lymphocyte proliferation immune responses in humans, Kramer and Burri[42] questioned whether these inconsistencies may be methodology related. The verdict is still out, and more work is needed to clarify the relationship between carotenes and other phytonutrients on immune response function.

A RECENT *IN VIVO* STUDY WITH RODENTS

Recently the inhibitory effect of four dietary carotenoids, α-carotene, β-carotene, lycopene, and lutein, prevalent in human blood and tissues against the formation of colonic aberrant crypt foci, has been reported in Sprague-Dawley rats.[43] The rats received 3 intrarectal doses of *N*-methylnitrosourea in week 1 and a daily gavage of de-escalated doses of carotenoids during weeks 2 and 5. Lycopene, lutein, α-carotene, and palm carotenes (a mixture of α-carotene, β-carotene, and lycopene) inhibited the development of aberrant crypt foci quantitated at week 6. In contrast, β-carotene was not effective. This study suggests that lycopene and lutein in small doses may potentially lower the risk for colon carcinogenesis.

BIOAVAILABILITY OF LUTEIN AND ZEAXANTHIN FROM PURIFIED SUPPLEMENTS

For the first time, Khachik et al. conducted a series of human bioavailability and metabolic studies with lutein and zeaxanthin. In the first study, 3 healthy Caucasian males (nonsmokers) between the ages of 42 and 59 ingested daily 10 mg of purified lutein (dispersed in olive oil) for 18 days.[21] The self-selected diet of the subjects largely excluded green, yellow, and orange fruits and vegetables containing lutein (i.e., broccoli, spinach, kale, green beans, green peas, lettuce, pumpkin, squash, peaches, oranges, and citrus fruits). The subjects kept daily dietary records throughout the study.

Baseline levels of plasma carotenoids were determined 25 and 11 days prior to lutein dosage (3 data points, days -25, -11, and 0). Plasma carotenoid profiles of the subjects were determined on days 0, 2, 4, 7, 14, 18, 26, 33, 40, and 57 using HPLC methodology. The blood levels of lutein in all three subjects increased 4- to 5-fold 1 week postsupplementation. In one subject, lutein plasma baseline level increased from ≈ 16 μg/dL (0.28×10^{-6} mol/L) to about 64 μg/dL (1.12×10^{-6} mol/L), resulting in the maximum absorption of this compound 1 week postsupplementation. The plasma concentrations of the lutein oxidation products (compounds 26–29 and 32, Table 1) increased significantly as the study progressed, indicating that the *in vivo* oxidation of lutein could be one of the key reactions in the human metabolism of this carotenoid.

In a second study designed similarly to the above, one subject ingested 20 mg/day of lutein dispersed in olive oil for 21 days. Plasma carotenoid profile of the subject was monitored on days 0, 2, 4, 7, 9, 11, 15, 18, 21, 25, 31, and 39. The plasma levels of lutein increased by 9-fold from 12 μg/dL (0.21 μmol/L) to 108 μg/dL (1.89 μmol/L) during 3 weeks of treatment.[8] After 21 days, the subject underwent a thorough eye exam at the National Eye Institute (NIH, Bethesda, MD), which revealed no unusual accumulation in the retina or ocular toxicity. The plasma levels of the oxidative metabolites of lutein (compounds 26–29, Table 1) increased by 2- to 3-fold.

In a similar study with zeaxanthin, which was isolated and purified from a Chinese fruit *(Lycium Chinese Mill)*, 3 subjects ingested daily supplements containing 10 mg of purified zeaxanthin (prepared as above) for 3 weeks; the carotenoid plasma levels were determined as above.[21] The plasma levels of zeaxanthin in all 3 subjects increased by 4-fold after 1 week of supplementation. In addition the plasma levels of lutein, 3′-epilutein, and the oxidation products of these carotenoids (compounds 26–29, Table 1) increased significantly.

Kostic et al administered lutein and β-carotene as single equimolar doses (0.5 μmol/kg body weight) to 8 adult subjects.[44] The subjects included 4 females and 4 males, aged 24 to 48 years, with body weights between 53 and 91 kg. Based on the weight of the subjects, the single doses of lutein and β-carotene were approximately 14 to 26 mg/day. After single doses, 13 blood samples were taken during the subsequent 35 days. The mean serum concentration of lutein showed a single maximum at 16 h post ingestion, while that of β-carotene peaked at 6 h and again at 32 h. Lutein and β-carotene cleared from the serum at approximately the same rate. The investigators concluded that carotenoids interact with each other during intestinal absorption, metabolism, and serum clearance. However, the metabolism of lutein and pharmacokinetics of its oxidative metabolites were not addressed. This, in part, may be due to the serum concentrations of the oxidative metabolites of lutein being too low (1–5 μg/dL) to be affected by a single-dose ingestion of this compound.

METABOLIC TRANSFORMATION OF CAROTENOIDS IN HUMANS

As stated earlier, among the non-vitamin A-active carotenoids, only lutein and lycopene appear to undergo metabolic transformations. Dietary (3R,3'R,6'R)-lutein and (3R,3'R)-zeaxanthin may be interconverted through a series of oxidation, reduction, and double-bond isomerization reactions, as shown in Figure 3. Once 3-hydroxy-β,ϵ-caroten-3'-one is formed, it may undergo double-bond isomerization to form 3'-hydroxy-ϵ,ϵ-caroten-3-one (compound 26, Figure 1); upon allylic oxidation, this compound may be converted to ϵ,ϵ-caroten-3,3'-dione (compound 29, Figure 1). 3'-Hydroxy-ϵ,ϵ-caroten-3-one and ϵ,ϵ-caroten-3,3'-dione may also be formed from direct allylic oxidation of lactucaxanthin (compound 13, Figure 1). However, the low serum concentration of lactucaxanthin, which is reflective of its limited dietary source, is also expected to influence the extent to which this compound undergoes oxidation. For a complete review of the metabolic transformation of lutein and zeaxanthin, see Khachik et al.[8,21] The human bioavailability and metabolic studies with purified supplements of lutein and zeaxanthin described earlier are in agreement with these metabolic reactions in humans.

The metabolism of lycopene is markedly more complicated because this compound is first oxidized at the 5,6-position to form lycopene 5,6-epoxide, which is extremely unstable, and undergoes cyclization to give an epimeric mixture of 2,6-cyclolycopene-1,5-epoxide A and B. Although these epoxides have not been detected in human serum, their corresponding epimeric diols, 2,6-cyclolycopene-1,5-diols A and B (compounds 33 and 34, Figure 1) are present. These diols may be formed from acidic and/or enzymatic ring opening of their respective epoxides. The structure of the diols in human serum has been confirmed by comparison of their HPLC-UV-VIS-MS profile with those of synthetic compounds.[14] The metabolites of lycopene identified in human serum consist of a novel 5-membered ring end group with three asymmetric centers at C-2, C-5, and C-6. Currently, while the relative configurations at C-2, C-5, and C-6 for synthetic 2,6-cyclolycopene-1,5-diol A and B have been determined by ^{1}H and ^{13}C-nuclear magnetic resonance, the absolute configurations of these diols are not known. The origin of the metabolites of lycopene in human serum may be due to the presence of these compounds in tomato-based products.[20] However, the concentration of 2,6-cyclolycopene-1,5-diols A and B in raw tomatoes and tomato-based products is extremely low and probably would not account for their presence in human serum.[20]

CONCLUSION

The majority of epidemiological studies to date has associated the high consumption of carotenoid-rich fruits and vegetables to a lower risk for several human cancers. Unfortunately, the scientific community at large has prematurely assigned this protective effect to a single carotenoid, β-carotene, because of its vitamin A activity. During the past 14 years, we have reported that fruits and vegetables contain approximately 40 to 50 carotenoids, some of which are found at markedly greater concentrations than β-carotene. Among these, 34 carotenoids, including 13 stereoisomers and 8 metabolites, have been identified in human serum and breast milk. Furthermore, the major dietary carotenoids have also been detected in human tissues of liver, lung, breast, and cervix. Although the presence of a wide spectrum of carotenoids in human serum, tissues, and fluids does not unequivocally indicate their health benefits, per se, numerous *in vitro* and *in vivo* studies strongly suggest that carotenoids exhibit wide-ranging biological activities. These include antioxidant/anti-inflammatory properties, up-regulation/enhancement of cellular communication, and induction of the activity of detoxication enzymes. In addition, the *in vivo* antitumor activity of several carotenoids against colon carcinogenesis in a rodent model has been reported. Here, we have presented data in support of the photoprotective role of lutein and zeaxanthin in human retina, and a proposed mechanism of how these carotenoids could lower the risk for AMD is also advanced. In several bioavailability and metabolic studies involving carotenoids, we provided preliminary data that suggest that carotenoids such as lutein, zeaxanthin, and lycopene are readily susceptible to *in vivo* oxidation in humans and therefore may serve as scavengers of the reactive electrophiles. The antioxidant capability of carotenoids may provide protection of DNA against oxidative damage. Much of the definitive details concerning the biological activities, role, and function of carotenoids in humans remain unexplored due to the focus on β-carotene. Future studies should concentrate on elucidating the bioavailability, metabolism, function, interaction, and efficacy of the spectrum of dietary carotenoids as well as their metabolites.

NOMENCLATURE

For convenience, the trivial names of certain carotenoids have been used throughout this text. Only the common rather than correct systematic names for carotenoids have been presented in Table 1. For in-chain geometrical isomers of carotenoids, the terms *all-E* and *Z,* which refer to all-*trans* and *cis* isomers of carotenoids, respectively, should be used instead of the old nomenclature. However, since many readers are more familiar with the old nomenclature, we have used the terms all-*trans* and *cis* throughout this manuscript. The R and S symbols refer to those carotenoids with known configurations. For several of the oxidative metabolites of lutein and zeaxanthin, the R and S symbols have not been used because the absolute configurations of these carotenoids with two or more centers of chirality are not known at present.

ACKNOWLEDGEMENT

The authors wish to thank Dr. James Cecil Smith, Jr., and Dr. Tim Kramer at the Beltsville Human Nutrition Research Center (U.S. Department of Agriculture, Beltsville, MD) for their helpful comments and discussions in preparation of this manuscript.

REFERENCES

1. Pfander, H., Eds: Gerspacher, M., Rychener, M., Schwabe, R. 1987. *Key to Carotenoids.* Basel, Birkhäuser, pp. 11–218.
2. Khachik, F., Beecher, G. R., Goli, M. B., Lusby, W. R. 1991. Separation, Identification, and Quantification of Carotenoids in Fruits, Vegetables and Human Plasma by High Performance Liquid Chromatography. *Pure Appl. Chem.,* 63(1):71–80.
3. Khachik, F., Beecher, G. R., Goli, M. B., Lusby, W. R., Ed: Packer, L. 1992. Separation and Quantification of Carotenoids in Foods. In *Methods in Enzymology,* New York, Academic Press, Vol. 213A, pp. 347–359.
4. Khachik, F., Beecher, G. R., Goli, M. B., Lusby, W. R., Smith, J. C., Jr. 1992. Separation and Identification of Carotenoids and Their Oxidation Products in Extracts of Human Plasma. *Anal. Chem.,* 64(18):2111–2122.
5. Khachik, F., Beecher, G. R., Goli, M. B., Lusby, W. R., Daitch, C. E., Ed: Packer, L. 1992. Separation and Quantification of Carotenoids in Human Plasma. In *Methods in Enzymology,* New York, Academic Press, Vol. 213A, pp. 205–219.
6. Khachik, F., Englert, G., Daitch, C. E., Beecher, G. R., Lusby, W. R., Tonucci, L. H. 1992. Isolation and Structural Elucidation of the Geometrical Isomers of Lutein and Zeaxanthin in Extracts from Human Plasma. *J. Chromatogr. Biomed. Appl.,* 582:153–166.
7. Khachik, F., Englert, G., Beecher, G. R. 1995. Isolation, Structural Elucidation, and Partial Synthesis of Lutein Dehydration Products in Extracts from Human Plasma. *J. Chromatogr. Biomed. Appl.,* 670:219–233.
8. Khachik, F., Steck, A., Pfander, H. 1997. Bioavailability, Metabolism, and Possible Mechanism of Chemoprevention by Lutein and Lycopene in Humans. In: *Food Factors for Cancer Prevention,* Ohigashi, H., Osawa, T., Terao, J., Watanabe, S., Yoshikawa, T., eds. Tokyo, Springer-Verlag, pp. 542–547.
9. Micozzi, M. E. (Ed). 1989. *Nutrition and Cancer Prevention.* New York, Marcel Dekker, pp. 213–241.
10. van Poppel, G. 1993. Carotenoids and Cancer: An Update with Emphasis on Human Intervention Studies. *Eur. J. Cancer,* 29A:1335–1344.

11. Morris, D. L., Kritchevsky, S. B., Davis, C. E. 1994. Serum Carotenoids and Coronary Heart Disease. *J. Am. Med. Assoc.* 272:1439–1441.

12. Gaziano, J. M., Manson, J. E., Branch, L. G., Colditz, G. A., Willet, W. C., Buring, J. E. 1995. A Prospective Study of Carotenoids in Fruits and Vegetables and Decreased Cardiovascular Mortality in the Elderly. *Ann Epidemiol.* 5:255–260.

13. Seddon, J. M., Ajani, U. A., Sperduto, R. D., Hiller, R., Blair, N., Burton, T. C., Farber, M. D., Gragoudas, E. S., Haller, J., Miller, D. T., Yannuzzi, L. A., Willet, W. 1994. Dietary Carotenoids, Vitamin A, C, and E, and Advanced Age-Related Macular Degeneration. *J. Am. Med. Assoc.* 272:1413–1420.

14. Khachik, F., Spangler, C. J. Smith, J. C. Jr., Canfield, L. M., Steck, A., Pfander, H. 1997. Identification, Quantification, and Relative Concentrations of Carotenoids and Their Metabolites in Human Milk and Serum. *Anal. Chem.* 69:1873–1881.

15. Khachik, F., Nir, Z., Ausich, R. L. 1997. Distribution of Carotenoids in Fruits and Vegetables as a Criterion for the Selection of Appropriate Chemopreventive Agent. In: *Food Factors for Cancer Prevention,* Ohigashi, H., Osawa, T., Terao, J., Watanabe, S., Yoshikawa, T., eds. Tokyo, Springer-Verlag, pp. 204–208.

16. Khachik, F., Beecher, G. R., and Whittaker, N. F. 1986. Separation Identification and Quantification of the Major Carotenoid and Chlorophyll Constituents in the Extracts of Several Green Vegetables by Liquid Chromatography. *J. Agric. Food Chem.,* 34(4):603–616.

17. Khachik, F., and Beecher, G. R. 1988. Separation and Identification of Carotenoids and Carotenol Fatty Acid Esters in Some Squash Products by Liquid Chromatography: Quantification of Carotenoids and Related Esters by HPLC. *J. Agric. Food Chem.,* 36(5):929–937.

18. Khachik, F., Beecher, G. R., and Lusby, W. R. 1989. Separation, Identification, and Quantification of the Major Carotenoids in Extracts of Apricots, Peaches, Cantaloupe, and Pink Grapefruit by Liquid Chromatography. *J. Agric. Food Chem.,* 37(6):1465–1473.

19. Khachik, F., Goli, M. B., Beecher, G. R., Holden, J., Lusby, W. R., Tenorio, M. D., and Barrera, M. R. 1992. The Effect of Food Preparation on Qualitative and Quantitative Distribution of Major Carotenoid Constituents of Tomatoes and Several Green Vegetables. *J. Agric. Food Chem.,* 40(3):390–398.

20. Tonucci, L. H., Holden, J. M., Beecher, G. R., Khachik, F., Davis, C. S., and Mulokozi, G. 1995. Carotenoid Content of Thermally Processed Tomato-Based Food Products. *J. Agric. Food Chem.,* 43:579–586.

21. Khachik, F., Beecher, G. R., Smith, J. C., Jr. 1995. Lutein, Lycopene, and Their Oxidative Metabolites in Chemoprevention of Cancer. *J. Cellular Biochem.,* 22:236–246.

22. Clevidence, B. A., Khachik, F., Brown, E. D., Nair, P. P., Wiley, E. R., Prior, R. L., Cao, G., Morel, D. W., Stone, W., Gross, M., Kramer, T. R. 1997. Consumption of Carotenoid-Rich Vegetables: Oxidation, Immune Response and Bioavailability. In: *Antioxidant Methodology: In vivo and in vitro Concepts.* Aruoma, O. and Cuppett, S., eds. Am. Oil Chemists Soc. Press, Champaign, IL pp. 53–63.

23. Bone, R. A., Landrum, J. T., Tarsis, S. L. 1985. Preliminary Identification of the Human Macular Pigment. *Vision Res.* 25:1531–1535.

24. Bone, R. A., Landrum, J. T., Fernandez, L., Tarsis, S. L. 1988. Analysis of the Macular Pigment by HPLC: Retinal Distribution and Age Study. *Invest. Ophthalmol. Vis. Sci.* 29:843–849.

25. Bone, R. A., Landrum, J. T., Hime, G. W., Cains, A. 1993. Stereochemistry of the Human Macular Carotenoids. *Invest. Ophthalmol. Vis. Sci.* 34:2033–2040.

26. Khachik, F., Bernstein, P., Garland, D. L. 1997. Identification of Lutein and Zeaxanthin Oxidation Products in Human and Monkey Retinas. *Invest. Ophthalmol. Vis. Sci.,* 38:1802–1811.

27. Bernstein, P. S., Balashov, N. A., Tsong, E. D., Rando, R. R. 1997. Retinal Tubulin Binds Macular Carotenoids. *Invest. Ophthalmol. Vis. Sci.,* 38:167–175.

28. Kelloff, G. J., Boone, C. W., Steele, V. E., Crowell, J. A., Lubet, R., Sigman, C. C. 1994. Progress in Cancer Chemoprevention: Perspectives on Agent Selection and Short-Term Clinical Intervention Trials. *Cancer Res.* 54 (Suppl.):2015s–2024s.

29. Harris, C. 1991. Chemical and Physical Carcinogenesis: Advances and Perspectives for the 1990s. *Cancer Res.* 51 (Suppl.):5023s–5044s.

30. Talalay, P., Eds: Wattenberg, L., Lipkin, M., Boone, C. W., Kelloff, G. J. 1992. The Role of Enzyme Induction in Protection Against Carcinogenesis. In *Cancer Chemoprevention,* Boca Raton, Florida, CRC Press, pp. 469–478.

31. Bertram, J. S. 1994. The Chemoprevention of Cancer by Dietary Carotenoids: Studies in Mouse and Human Cells. *Pure Appl. Chem.,* 66:1025–1032.

32. King, T. J., Khachik, F., Bortkiewicz, H., Fukushima, L. H., Morioka, S., Bertram, J. S. 1997. Metabolites of Dietary Carotenoids as Potential Cancer Preventive Agents. *Pure Appl. Chem.,* 69:2135–2140.

33. Bendich, A., Olson, J. A. 1989. Biological Actions of Carotenoids. *FASEB J.* 3:1927–1932.

34. Baker, K. R., Meydani, M. 1993. β-Carotene as an Antioxidant in Immunity and Cancer. *J. Optimal Nutr.* 3:39–50.

35. Middleton, E. Jr., Kandaswami, C., Ed: Harborne, J. B. 1993. The Impact of Plant Flavonoids on Mammalian Biology: Implications for Immunity, Inflammation and Cancer. In *The Flavonoids: Advances in Research since 1986,* London, Chapman & Hall, pp. 619–652.

36. Berg, P. A., Daniel, P. T., Eds: Cody, V., Middleton, E., Jr., Harborne, J. B., Beretz, A. Effects of Flavonoid Compounds on the Immune Response. In *Plant Flavonoids in Biology and Medicine II: Biochemical, Cellular, and Medicinal Properties,* New York, Alan R. Liss, pp. 157–71.

37. Bendich, A., Shapiro, S. S. 1986. Effects of β-Carotene and Canthaxanthin on the Immune Responses of the Rat. *J. Nutr.* 116:2254–2262.

38. Bendich, A. 1991. β-Carotene and the Immune Response. *Proc. Nutr. Soc.* 50:263–274.

39. van Poppel, G., Spanhaak, S., Ockhuizen, T. 1993. Effect of β-Carotene on Immunological Indexes in Healthy Male Smokers. *Am. J. Clin. Nutr.* 57:402–407.

40. Moriguichi, S., Okishama, N., Sumida, S., Okamura, K., Doi, T., Kishino, Y. 1996. β-Carotene Supplementation Enhances Lymphocyte Proliferation with Mitogens in Human Peripheral Blood Lymphocytes. *Nutr. Res.* 16:211–218.

41. Daudu, P. A., Kelley, S. S., Taylor, P. C., Burri, B. J., Wu, M. M. 1994. Effect of a Low β-Carotene Diet on the Immune Functions of Adult Women. *Am. J. Clin. Nutr.* 60:969–72.

42. Kramer, T. R., Burri, B. J. 1997. Modulated Mitogenic Proliferative Responsiveness of Lymphocytes in Whole-Blood Cultures after a Low-Carotene Diet and Mixed-Carotenoid Supplementation in Women. *Am. J. Clin. Nutr.* 65:871–875.

43. Narisawa, T., Fukaura, Y., Hasebe, M., Ito, M., Aizawa, R., Murakoshi, M., Uemura, S., Khachik, F., and Nishino, H. 1996. Inhibitory Effects of Natural Carotenoids, α-Carotene, β-Carotene, Lycopene and Lutein, on Colonic Aberrant Crypt Foci Formation in Rats. *Cancer Letters,* 107:137–142.
44. Kostic, D., White, W. S., Olson, J. A. 1995. Intestinal Absorption, Serum Clearance, and Interactions Between Lutein and β-carotene When Administered to Human Adults in Separate or Combined Oral Doses. *Am. J. Clin. Nutr.,* 62:604–610.

CHAPTER 6

Carotenoid – Carotenoid Interactions

WENDY S. WHITE
INKE PAETAU

INTRODUCTION

CAROTENOIDS are considered to follow the general scheme for intestinal lipid absorption without specific mechanisms for uptake, intracellular transport, or incorporation into lipoproteins by enterocytes. However, mammalian species are differentiated according to the ability or inability to absorb carotenoids intact in the small intestine and are further subdivided according to the ability to accumulate primarily hydrocarbon carotenes or to indiscriminately accumulate carotenes and oxycarotenoids (xanthophylls) in blood and tissues (Goodwin, 1984). The species selectivity of intestinal carotenoid absorption suggests the evolution of intestinal absorption pathways that are selective for carotenoids, in general, and that discriminate between hydrocarbon carotenes and oxy-carotenoids. Humans are among only a few species that indiscriminately absorb both carotenes and oxycarotenoids in the small intestine. Our research findings suggest metabolic heterogeneity and specific interactions of these two classes of carotenoids during intestinal absorption in humans.

INTESTINAL METABOLISM AND ABSORPTION OF CAROTENOIDS

Intestinal absorption of carotenoids from plant foods is limited by food matrix effects, whereas, for purified carotenoids, micellar solubilization and uptake by intestinal cells are more likely to be limiting (Parker, 1996). Little is known regarding the micellar solubilization of carotenoids in the intestinal lumen as a vehicle for uptake into intestinal epithelial cells. Micellar solubilization markedly increases the rate of β-carotene uptake relative to emulsions, and uptake is further enhanced in the presence of bile salts (El-Gorab et al., 1975). A recent study is informative regarding the assumed transfer of carotenoids from dietary emulsions to mixed micelles in the intestinal lumen (Borel et al., 1996). In phospholipid-stabilized triacylglycerol emulsions, β-carotene distributed almost exclusively in the triacylglycerol core, whereas a model oxycarotenoid, zeaxanthin, distributed preferentially in the surface phospholipids. This suggests a greater spontaneous transfer of polar oxycarotenoids, such as lutein and zeaxanthin, from dietary emulsions to mixed micelles and a potential difference in the micellar solubilization of these carotenoids and hydrocarbon carotenes, such as α-carotene, β-carotene, and lycopene.

The unstirred water layer could be a barrier to uptake of carotenoids by enterocytes if diffusion through the unstirred water layer is rate limiting relative to passage across the lipid bilayer of the brush-border membrane (Hollander and Ruble, 1978). There is evidence that a low pH microclimate may favor the diffusion and dissociation of mixed micelles across the unstirred water layer (Shiau and Levine, 1980). This may account for the adverse effect of high intraluminal pH on intestinal β-carotene absorption. In humans, achlorhydria induced pharmacologically by omeprazole inhibits intestinal β-carotene absorption, as indicated by a reduced serum increment in response to an oral β-carotene dose (Tang et al., 1996). In rats, an increase in hydrogen ion concentration enhances the rate of uptake of β-carotene from a micellar perfusate into intestinal mucosal cells (Hollander and Ruble, 1978). Thus, atrophic gastritis, a common condition in older adults characterized by reduced gastric acid secretion, may contribute to the wide interindividual variation in the extent of absorption of intact β-carotene (Tang et al., 1996).

Intestinal perfusion studies in rats indicate that the mechanism of uptake of all-*trans* β-carotene at the brush-border membrane of the small intestine is passive diffusion (Hollander and Ruble, 1978). There is a linear relation between the perfusate concentration of [15-15′-^{14}C] β-carotene and its disappearance rate from the perfusate, which is not consistent with a capacity-limited or carrier-mediated mechanism. The time and concentration dependence of β-carotene uptake into rat brush-border membrane vesicles (El-Gorab et al., 1975) or a rat small intestinal epithelial cell line (Scita et al., 1992) also indicate a diffusion-mediated mechanism.

The mechanisms of intracellular transport of absorbed carotenoids and incorporation into intestinal lipoproteins are unknown. Intracellular transport might potentially involve cytosolic lipid-binding proteins analogous to phosphatidylcholine transfer protein (PC-TP), which binds one PC molecule for transfer between membranes (Wirtz and Gadella, 1990), and intestinal-type cytosolic fatty acid binding protein (I-FABP$_c$), which transports long-chain fatty acids to the endoplasmic reticulum (Ockner and Manning, 1974). However, Gugger and Erdman (1996) did not observe cytosolic protein-mediated β-carotene transfer when liposomes containing [^{3}H] β-carotene were incubated with isolated bovine hepatic mitochondria and increasing quantities of intestinal mucosal cell cytosol. Mitochondria were used as the acceptor membrane system, based on established protocols for transfer assays for cholesterol, phospholipids, and α-tocopherol. It was concluded that intracellular β-carotene transport is mediated by mechanisms other than by cytosolic transport proteins such as by vesicular transport or membrane-bound proteins.

A familiar example of the specificity of intestinal carotenoid metabolism is the selective conversion of carotenoids with a molecular structure characterized by at least one unsubstituted β-ionone ring to vitamin A. This indicates that the interaction of the β-ionone ring with the cleavage enzyme is important in catalysis (Ershov et al., 1994). The presence of two unsubstituted β-ionone rings in the molecular structure of β-carotene predicts its greater biopotency relative to other provitamin A carotenoids. The biopotency of other prominent dietary provitamin A carotenoids, α-carotene and β-cryptoxanthin, is approximately half that of β-carotene in rats as predicted by the presence of only one unsubstituted β-ionone ring in these molecules (Zechmeister, 1962). The conversion efficiency of provitamin A carotenoids is also stereospecific; with rare exceptions, the *cis* forms have lower biopotency than the respective all-*trans* carotenoids. However, the determinant of enzymatic cleavage is most likely the conformation of the enzyme substrate complex rather than the geometric configuration of the carotenoid per se (Zechmeister, 1962). For example, all-*trans*-γ-carotene and a poly*cis* form have equal provitamin A activities in rats (Zechmeister, 1962); in chicks the poly*cis* isomer has even greater provitamin A activity than the all-*trans* form (Greenberg et al., 1949).

Biostereoisomerization of carotenoids in the gastrointestinal tract was reported in rats and chicks in the, 1940s (Kemmerer and Fraps, 1945; Deuel et al., 1951). More recently, You et al. (1996) showed that a significant proportion of a tracer dose of [^{13}C] 9-*cis*-β-carotene is isomerized to [^{13}C] all-*trans*-β-carotene before entering the circulation in human subjects. Thus, bioisomerization to [^{13}C] all-*trans*-β-carotene contributes to the documented near absence of [^{13}C] 9-*cis*-β-carotene in human plasma (Jensen et al., 1987; Stahl et al., 1993) or plasma chylomicrons after ingestion of an oral dose (Stahl et al., 1993). The findings of You et al. (1996) further indicate that isomerization to all-*trans*-β-carotene contributes to the provitamin A activity of 9-*cis*-β-carotene.

Central cleavage of β-carotene to vitamin A by β-carotene 15-15′-dioxygenase was first shown *in vivo* by Olson (1961) and then independently *in vitro* by Olson and Hayaishi (1965) and Goodman and Huang (1965). There is recent extensive evidence that β-carotene is also metabolized by excentric cleavage of the double-bond chain to produce retinoic acid and a series of intermediate β-apo-carotenals (Wolf, 1995). For example, retinoic acid is formed from β-carotene added to an intestinal perfusate in ferrets in the presence or absence of citral, an inhibitor of retinaldehyde oxidation (HÈbuterne et al., 1996). Because retinaldehyde is an intermediate for the formation of retinoic acid via central cleavage, this study provides evidence for excentric cleavage. There may be intracellular compartmentalization of the two cleavage activities (Dimitrovskii, 1991). However, in a recent study of the subcellular localization of β-carotene cleavage activity in rat enterocytes, β-carotene cleavage activity was associated predominantly with the cytosolic fraction and only minimal activity with the microsomal and mitochondrial fractions (Duszka et al., 1996). The only detected metabolites were retinal and retinoic acid; the intermediate β-apo-carotenals of excentric cleavage were not detected in any subcellular fraction. The results suggest that the major conversion pathway for the formation of vitamin A from β-carotene in the rat intestine is central cleavage.

Whereas the mechanism of intracellular transport of carotenoids is unknown, cytoplasmic retinol binding proteins, CRBP and CRBP II, that have important roles in retinoid transport and metabolism have been isolated and well characterized. The localization of CRBP II within the enterocytes of the proximal intestinal epithelium suggests an important role in vitamin A absorption and/or esterification (Ong and Page, 1987; Crow and Ong, 1985). A role for CRBP II in carotenoid metabolism is also proposed. Retinal, the oxidative cleavage product of provitamin A carotenoids, is reduced to retinol prior to esterification. Both retinal and retinol bind CRBP II with high affinities (MacDonald and Ong, 1987). Retinal bound to CRBP II is a substrate for a microsomal retinal reductase. The product of this reaction, retinol bound to CRBP II, is subsequently esterified by lecithin-retinol acyltransferase and incorporated into chylomicrons. Conversion of provitamin carotenoids into vitamin A might potentially be regulated by requiring apoCRBP II for release of product inhibition (Ong, 1993).

The assembly and exocytosis of lipoproteins is the rate-limiting step for transport of lipids (Thomson et al., 1993), and, most likely, carotenoids, into the lymph. The intestine secretes two major lipoproteins, chylomicrons and very-low density lipoproteins (VLDLs), which appear to have divergent assembly pathways as indicated by Pluronic L-81 inhibition of the formation of chylomicrons but not of VLDL (Nutting et al., 1989). There are modifications of the phospholipids and proteins of chylomicrons and VLDL during maturation and transfer from the endoplasmic reticulum to the Golgi, which are the site of exocytosis (Thomson et al., 1993). The mechanisms of incorporation of carotenoids during intracellular assembly and secretion of intestinal lipoproteins across the basolateral membrane are unknown.

POSTPRANDIAL KINETICS OF β-CAROTENE AND OXYCAROTENOIDS

Our recent findings suggest distinct patterns of incorporation of β-carotene and oxycarotenoids into postprandial plasma lipoproteins in humans (Paetau et al., in press). The individual appearances of β-carotene and a model oxycarotenoid, canthaxanthin, in plasma and plasma lipoproteins were investigated and compared within normolipidemic premenopausal women (n = 9) after ingestion of a single or combined dose of each carotenoid (47 μmol β-carotene and/or 44 μmol canthaxanthin) during each of three study periods. The order of the carotenoid treatments was randomly assigned, and doses were separated by washout periods of 10 weeks to minimize residual effects of the preceding dose. After collection of a baseline blood sample, the β-carotene and/or canthaxanthin dose was ingested with a standardized fat-rich meal. Blood samples were collected hourly for 12 h postdosing via an indwelling intravenous catheter inserted in a forearm vein; additional blood samples were collected via venipuncture after an overnight fast at selected intervals for 24 to 528 hrs postdosing. In a subset of the women (n = 5), plasma collected at 0, 2, 4, 6, 8, and 10 h was separated by cumulative rate ultracentrifugation into chylomicrons, three VLDL subfractions of decreasing particle size and increasing density, intermediate-density lipoproteins (IDLs), and low-density lipoproteins (LDLs). The mean particle diameters determined by electron microscopy and computerized image analysis were 100.0 ± 0.0 nm, chylomicrons; 46.0 ± 2.4 nm, VLDLA; 37.8 ± 2.6 nm, VLDLB; 24.8 ± 3.3 nm, VLDLC; and 14.0 ± 0.1 nm, LDL (IDL was not measured).

The individual appearances of β-carotene and canthaxanthin were compared in plasma and plasma lipoproteins after each subject ingested a separate dose of each carotenoid (47 and 44 μmol, respectively). There was a preferential appearance of the oxycarotenoid canthaxanthin relative to that of β-carotene in plasma triacylglycerol-rich lipoproteins (TRLs) during the 10-h postprandial period investigated (Figures 1 and 2). The areas under the curves (AUCs) for the change from baseline as a function of time were greater for canthaxanthin than for β-carotene in chylomicrons ($P = 0.05$), VLDLA ($P = 0.02$), VLDLB ($P = 0.06$), and VLDLC ($P = 0.08$) (data not shown). A potential explanation for the difference is provitamin A activity; canthaxanthin is a non-provitamin A carotenoid in mammals. However, intestinal conversion to retinoid metabolites is not likely to account for the extent of reduced appearance of β-carotene in TRLs relative to that of canthaxanthin; the cleavage enzyme(s) is homeostatically regulated and the efficiency of the cleavage reaction would be inhibited by ingestion of a 47-μmol dose of purified β-carotene. This conclusion is supported by comparison of the total plasma response from 0 to 96 h postdosing, which includes both the postprandial and postabsorptive periods (Table 1); the 0- to 96-h plasma AUCs for canthaxanthin and β-carotene were similar

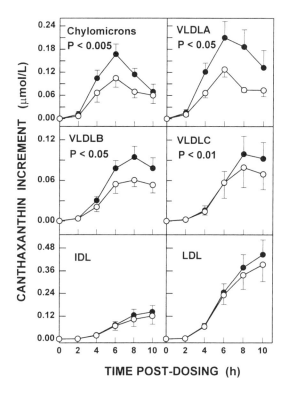

Figure 1 Changes of canthaxanthin from baseline in plasma chylomicrons, VLDL subfractions A, B, and C, IDL, and LDL after each subject ingested canthaxanthin (44 μmol) (●) and β-carotene (47 μmol) plus canthaxanthin (44 μmol) (○) separated by a washout period of at least 10 weeks. Data represent mean values for 5 subjects ± SEM; data for LDL canthaxanthin are missing for one subject. The P values shown correspond to the effect of canthaxanthin versus β-carotene plus canthaxanthin analyzed by repeated measures analysis of variance (ANOVA). (Reproduced from *Am. J. Clin. Nutr.* 1997;66:1133–1143. © American Society for Clinical Nutrition.)

($P = 0.62$), whereas the 0 to 10 h plasma AUCs were significantly different ($P < 0.0001$) (data not shown). Our findings are consistent with those of Gärtner et al. (1996), who compared the increments of the hydrocarbon carotenes, α-carotene and β-carotene, and those of the oxycarotenoids, luteinand zeaxanthin, in the plasma chylomicron fractions of human subjects 0 to 12 h after ingestion of a mixed carotenoid dose in the form of an algal extract, Betatene. There was a preferential increase of the oxycarotenoids lutein and zeaxanthin, in chylomicrons relative to that of β-carotene, as indicated by the carotenoid composition of chylomicrons relative to that in the dose. In contrast, the appearance of α-carotene in chylomicrons was consistent with the α-carotene content of Betatene, a natural carotenoid raw material.

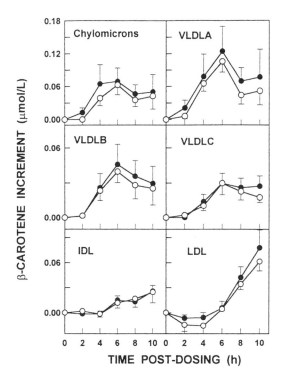

Figure 2 Changes of β-carotene from baseline in plasma chylomicrons, VLDL subfractions A, B, and C, IDL, and LDL after each subject ingested β-carotene (47 μmol) (●) and β-carotene (47 μmol) plus canthaxanthin (44 μmol) (○) separated by a washout period of at least 10 weeks. Data represent mean values for 5 subjects ± SEM. (Reproduced from *Am. J. Clin. Nutr.* 1997;66:1133–1143. © American Society for Clinical Nutrition.)

The postprandial appearances of β-carotene and canthaxanthin in large VLDL (VLDLA) in our study exceeded those in other TRL subfractions (Figures 1 and 2). This is consistent with an earlier human study in which the total contribution of VLDL to the plasma β-carotene increment was greater than that of chylomicrons (Cornwell et al., 1962). Increases of both apolipoprotein (apo) B-48, the structural protein associated with chylomicrons, and apo B-100 in TRLs were reported after ingestion of either fat-rich or conventional meals by human subjects (Cohn et al., 1988; Karpe et al., 1993; Schneeman et al., 1993). The predominant lipoprotein particle accumulated in postprandial lipoproteinemia is VLDL rather than chylomicrons; 80% of the increase in TRL particle number is attributed to VLDL containing apo B-100 and only 20% to chylomicrons containing apo B-48 (Schneeman et al., 1993). This postprandial increase of apo B-100 concentration is confined to the large VLDL subfraction (Karpe et al., 1993, 1995). Thus the marked

TABLE 1. Within-Subject Comparison of the Areas under the Plasma Concentration-Time Curves (AUC) for 0–96 h after Ingestion of an Individual or Combined Dose of β-Carotene (47 μmol) and Canthaxanthin (44 μmol).

Subject	Area under Curve (mmol • h/L)					
	β-Carotene plus Placebo	β-Carotene plus Canthaxanthin	Within Subject Difference	Canthaxanthin plus Placebo	Canthaxanthin Plus β-Carotene	Within Subject Difference
1	36.2	40.4	+4.2	98.1	87.1	−11.0
2	113.0	90.0	−23.0	79.9	76.9	−3.0
3	149.0	96.1	−52.9	79.4	65.8	−13.6
4	2.6	26.7	+24.1	46.4	32.2	−14.2
5	23.1	34.3	+11.2	51.1	38.5	−12.6
6	47.9	12.4	−35.5	62.4	53.2	−9.2
7	34.6	57.7	+23.1	78.2	77.0	−1.2
8	102.3	53.0	−49.3	71.4	63.1	−8.3
9	65.0	46.3	−18.7	73.0	69.6	−3.4
Mean ± SEM	63.7 ± 16.0	50.8 ± 9.2	−13.0 ± 9.9	71.1 ± 5.3	62.6 ± 6.1	−8.5 ± 1.6

Reproduced from Am. J. Clin. Nutr. 1997 66:1133–1143. © American Society for Clinical Nutrition.

appearance of β-carotene and canthaxanthin in large VLDL in our study may reflect an early postprandial increase in particle number; the marked appearance of canthaxanthin would further suggest an intestinal rather than hepatic origin for the VLDL particles because endogenous canthaxanthin concentrations are low.

There was a rapid accumulation of the oxycarotenoid canthaxanthin in LDL relative to that of the hydrocarbon carotene β-carotene during the immediate postprandial period. Within 6 h postdosing, the change of canthaxanthin from baseline in LDL exceeded that in chylomicrons and other TRLs (Figure 1). At 10 h postdosing, the AUCs for canthaxanthin were significantly greater than those for β-carotene in LDL ($P = 0.007$) (data not shown); at 10 h postdosing, the percentages of the increases of plasma canthaxanthin and β-carotene that were associated with LDL were 32.4 ± 3.6 % and, $19.8 \pm 3.3\%$, respectively ($P < 0.05$). Preferential accumulations in TRLs and LDL contribute to the greater plasma response to ingested canthaxanthin relative to ingested β-carotene during the immediate postprandial period.

The plasma appearance curves of β-carotene and canthaxanthin are distinct (Figure 3). Plasma β-carotene concentrations peak initially at 5 h, with a larger

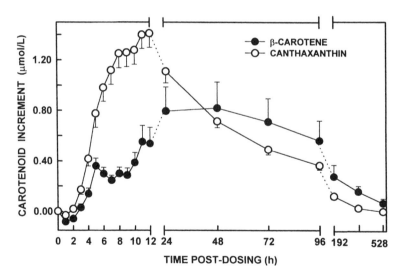

Figure 3 Changes of plasma β-carotene and canthaxanthin concentrations from baseline after each subject ingested β-carotene (47 μmol) and canthaxanthin (44 μmol) separated by a washout period of at least 10 weeks. Data represent mean values for 9 subjects ± SEM. The main effect of β-carotene versus canthaxanthin ($P = 0.\ 0001$), the main effect of time ($P = 0.\ 0001$), and the interaction (β-carotene versus canthaxanthin × time) ($P = 0.\ 0001$) were significant by repeated measures analysis of variance (ANOVA). (Reproduced from *Am. J. Clin. Nutr.* 1997; 66:1133–1143. © American Society for Clinical Nutrition.)

sustained peak from 24 to 48 h postdosing. Our findings (Paetau et al., in press) and those of others (Cornwell et al., 1962; Johnson and Russell, 1992) indicate that the initial peak coincides with β-carotene increments in TRLs and the major peak with delayed appearance of β-carotene in LDL. The interval that separates the two peaks is presumed to reflect uptake of β-carotene in chylomicron remnants by the liver, incorporation of β-carotene into nascent VLDL secreted by the liver, and subsequent metabolism to LDL. In contrast, the plasma appearance of the oxycarotenoids canthaxanthin (White et al., 1994; Paetau et al., in press) and lutein (Kostic et al., 1995) is monophasic, with a single peak at 12 to 24 h postdosing. Our recent findings indicate that the monophasic plasma appearance of the model oxycarotenoid canthaxanthin reflects coincident increments in plasma TRL and LDL (Paetau et al., in press).

POSTPRANDIAL INTERACTIONS OF β-CAROTENE AND OXYCAROTENOIDS

Interactions of the postprandial appearance of β-carotene and a model oxycarotenoid, canthaxanthin, in plasma and plasma TRLs were investigated after concurrent ingestion by human subjects (Paetau et al., in press). As described previously, normolipidemic women ingested either an individual (47-μmol) β-carotene dose, an individual (44-μmol) canthaxanthin dose, or a combined β-carotene and canthaxanthin dose during each of three study periods. The kinetics of the plasma appearance of the individual and combined doses of the carotenoids were investigated and compared within subjects. Repeated measures of analysis of variance indicated that ingestion of the combined dose of β-carotene and canthaxanthin significantly inhibited the appearance of canthaxanthin but not that of β-carotene in plasma and plasma TRLs (Figures 1 and 2). The AUCs for canthaxanthin were significantly reduced in chylomicrons by $37.8 \pm 7.1\%$ ($P < 0.005$), in VLDLA by $43.0 \pm 8.1\%$ ($P < 0.05$), in VLDLB by $32.7 \pm 8.1\%$ ($P < 0.05$), and in VLDLC by $30.3 \pm 7.6\%$ ($P < 0.005$) (data not shown). Ingestion of the combined dose of β-carotene and canthaxanthin did not significantly inhibit the rapid accumulation of canthaxanthin in LDL during the immediate 10-h postprandial period.

Our findings are consistent with the pilot study (White et al., 1994) and indicate a specific interaction of β-carotene and the oxycarotenoid canthaxanthin during intestinal absorption. Our findings are also supported by a recent investigation of interactive effects of concurrent ingestion of β-carotene and the oxycarotenoid lutein (Kostic et al., 1995). Ingestion of a combined equimolar dose of the two carotenoids significantly inhibited the appearance of lutein but not that of β-carotene in serum. In contrast, Johnson et al. (1996) did not observe an inhibition of the serum appearance of either β-carotene or lycopene, two hydrocarbon carotenoids, after ingestion of a combined as compared with an individual dose. These studies support the hypothesis that carotenoid-carotenoid interactions during intestinal absorption are limited to interactions of β-carotene (or possibly other hydrocarbon carotenes) and oxy-carotenoids (White et al., 1994).

Coincident ingestion of the β-carotene dose reduced the plasma canthax-anthin AUC resulting from ingestion of the canthaxanthin dose in each of nine subjects (Table 1). In contrast, there was no consistent effect of coincident ingestion of the canthaxanthin dose on the plasma β-carotene AUC resulting from ingestion of the β-carotene dose. In five of nine subjects, the plasma β-carotene AUC decreased, and, in the remaining four subjects, the plasma β-carotene increment increased. These four subjects were those who had the lowest AUCs after ingestion of the individual β-carotene dose. In a separate study, coincident ingestion of the oxycarotenoid lutein also enhanced the serum β-carotene increments of those subjects who had the lowest serum AUCs after ingestion of β-carotene (Kostic et al., 1995). It was hypothesized that subjects who have low serum β-carotene increments in response to ingestion of an oral dose ("nonresponders") might metabolize β-carotene efficiently to retinoids in intestinal epithelial cells. In those individuals, non-provitamin A oxycarotenoids might form a pseudosubstrate complex and inhibit the cleavage enzyme(s) so that more β-carotene is absorbed intact. The effect of non-provitamin A carotenoids on the β-carotene 15,15′-dioxy-genase assay is a current research focus. Conversion of β-carotene to retinal in hamster intestinal preparations is reported to be substantially inhibited by incubation with the non-provitamin A oxycarotenoid lutein but not by incubation with the non-provitamin A hydrocarbon carotene lycopene (van Vliet et al., 1996). However, a similar study of the β-carotene 15,15′-dioxygenase assay found that addition of lycopene to the incubation had a more pronounced inhibitory effect than addition of lutein (Ershov et al., 1994).

HEALTH IMPLICATIONS

Our findings are limited to interactions of β-carotene and the oxy-carotenoids canthaxanthin and lutein occurring during intestinal absorption and after coincident ingestion of purified doses. Such investigations are useful probes of the mechanism(s) of intestinal carotenoid absorption; however, the health implications of the demonstrated carotenoid–carotenoid interactions remain to be determined. The findings of major clinical trials have focused concern on potential metabolic imbalances created by chronic β-carotene supplementation, including competition with other dietary carotenoids, which are potentially health protective, for intestinal absorption. In separate large-scale randomized trials, the Alpha-Tocopherol, Beta-Carotene (ATBC) Cancer Prevention Study and the Beta-Carotene and Retinol Efficacy Trial (CARET), supplementation with β-carotene alone or in combination with retinyl palmitate was associated with apparent promotion of lung carcinogenesis in current cigarette smokers (The Alpha-Tocopherol, Beta Carotene Cancer Prevention Study Group, 1994; Omenn et al., 1996b). In a third clinical trial, the Physicians Health Study, β-carotene supplementation resulted in neither benefit nor harm in a population of healthy male physicians at low risk for lung cancer on the basis of smoking history (Hennekens et al., 1996).

A potential biological mechanism for the increased risks of lung cancer in cigarette smokers in the ATBC and CARET trials is an interaction of β-carotene with ethanol (Leo and Lieber, 1994); excess lung cancer incidences in the intervention groups were associated with those heavy smokers with the highest levels of alcohol intakes (Omenn et al., 1996a; Albanes et al., 1996). However, the absence of a dose-response effect with alcohol intake reduces the plausibility of this mechanism (Omenn et al., 1996a). The high serum β-carotene concentrations achieved in these studies might also theoretically be associated with a pro-oxidative interaction of β-carotene with cigarette smoke in the lung (Albanes et al., 1996). Antagonistic effects of supplemental β-carotene on the intestinal absorption of protective micronutrients or phytochemicals, including other carotenoids, are another potential mechanism underlying the observed increase in lung cancer risk (Albanes et al., 1996).

There are limited and inconsistent data regarding effects of chronic β-carotene supplementation on the bioavailability of carotenoids ingested from the diet. In a human metabolic study, men who consumed a controlled low-carotenoid diet for 6 weeks and ingested daily either 12 or 30 mg of purified β-carotene had significantly lower plasma lutein concentrations relative to men who consumed the same diet and ingested a placebo (Micozzi et al., 1992). The dosing schedule was conducive to interactions with dietary carotenoids; purified β-carotene doses were consumed with meals and were split between lunch and dinner. In contrast, initial results of the Physicians Health Study are not consistent with adverse effects of β-carotene supplementation on plasma carotenoid concentrations in free-living populations. In a small sample of participants, there was no effect of approximately 12 years of β-carotene supplementation on plasma concentrations of other carotenoids, including the oxycarotenoid lutein and cryptoxanthin (Fotouhi et al., 1996). Because β-carotene supplements were ingested on alternate days rather than daily, the potential for interactions with dietary carotenoids would be expected to be minimized. In the Australian Polyp Prevention Study, significant increases in serum concentrations of the hydrocarbon carotenes α-carotene and lycopene were reported after 24 months of daily β-carotene supplementation, and there was no significant change of serum concentrations of lutein from baseline (Wahlqvist et al., 1994). In these clinical trials, ingestion of β-carotene supplements in relation to ingestion of meals or carotenoid-rich foods was not standardized. In this respect, well-controlled human metabolic studies are more likely to be conclusive regarding adverse interactions of β-carotene supplements and dietary oxycarotenoids during coincident ingestion.

CONCLUSIONS

Our research findings indicate specific interactions of β-carotene and the oxycarotenoids canthaxanthin and lutein during intestinal absorption. In each human study, the interactions were not reciprocal; ingestion of a combined equimolar dose of β-carotene and oxycarotenoid significantly inhibited the appearance of oxycarotenoid but not that of β-carotene in plasma. These findings suggest (1) specific and distinct mechanisms of intestinal absorption for β-carotene and oxycarotenoids and (2) that β-carotene supplements may inhibit intestinal absorption of dietary oxycarotenoids, which are efficient antioxidants. Whether β-carotene supplementation has a significant effect on blood concentrations or metabolism of oxycarotenoids in free-living populations is under investigation.

ACKNOWLEDGEMENTS

This research was supported by USDA grant 94-34115-0269 to the Center for Designing Foods to Improve Nutrition, Iowa State University, Journal Paper No. J-17493 of the Iowa Agriculture and Home Economics Experiment Station, Ames, Iowa, Project No. 3171, and supported by Hatch Act and State of Iowa funds.

REFERENCES

Albanes, D., Heinonen, O. P., Taylor, P. R., Virtamo, J., Edwards, B. K., Rautalahti, M., Hartman, A. M., Palmgren, J., Freedman, J. S., Haapakoski, J., Barrett, M. J., Pietinen, P., Malila, N., Tala, E., Liippo, K., Salomaa, E.-R., Tangrea, J. A., Teppo, L., Askin, F. B., Taskinen, E., Erozan, Y., Greenwald, P., and Huttunen, J. K., 1996. α-Tocopherol and β-Carotene Supplements and Lung Cancer Incidence in the Alpha-Tocopherol, Beta-Carotene Cancer Prevention Study: Effects of Baseline Characteristics and Study Compliance," *J. Natl. Cancer Inst.,* 88:1560–1570.

The Alpha-Tocopherol, Beta Carotene Cancer Prevention Study Group., 1994. The Effect of Vitamin E and Beta-carotene on the Incidence of Lung Cancer and Other Cancers in Male Smokers, *N. Engl. J. Med.,* 330:1029–1035.

Borel, P., Grolier, P., Armand, M., Partier, A., Lafont, H., Lairon, D., and Azais-Braesco, V. 1996. Carotenoids in Biological Emulsions: Solubility, Surface-to-Core Distribution, and Release from Lipid Droplets, *J. Lipid Res.,* 37:250–261.

Cohn, J. S., McNamara, J. R., Cohn, S. D., Ordovas, J. M., and Schaefer, E. J., 1988. Plasma Apolipoprotein Changes in the Triglyceride-rich Lipoprotein Fraction of Human Subjects Fed a Fat-rich Meal, *J. Lipid Res.,* 29:925–936.

Cornwell, D. G., Kruger, F. A., and Robinson, H. B., 1962. Studies on the Absorption of Beta- Carotene and the Distribution of Total Carotenoid in Human Serum Lipoproteins after Oral Administration, *J. Lipid Res.,* 3:65–70.

Crow, A., and Ong, D. E., 1985. Cell-Specific Immunohistochemical Localization of a Cellular Retinol-Binding Protein (Type Two) in the Small Intestine of Rat, *Proc. Natl. Acad. Sci. USA,* 82:4707–4711.

Deuel, H. J., Jr., Ganguly, J., Koe, B. K., and Zechmeister, L., 1951. Stereoisomerization of the Poly*cis* Compounds, Pro-γ-Carotene and Prolycopene in Chickens and Hens, *Arch. Biochem. Biophys.*, 33:143–149.

Dimitrovskii, A. A., 1991. In: Ozawa, T., ed. *New Trends in Biological Chemistry.* Berlin: Springer-Verlag, pp. 297–308.

Duszka, C., Grolier, P., Azim, E. -M., Alexandre-Gouabau, M.-C., Borel, P., and Azais-Braesco, V., 1996. Rat Intestinal β-Carotene Dioxygenase Activity is Located Primarily in the Cytosol of Mature Jejunal Enterocytes, *J. Nutr.,* 126:2550–2556.

El-Gorab, M. I., Underwood, B. A., and Loerch, J. D., 1975. The Roles of Bile Salts in the Uptake of β-Carotene and Retinol by Rat Everted Gut Sacs, *Biochim. Biophys. Acta,* 401:265–277.

Ershov, Y. V., Bykhovsky, V. Y., and Dmitrovskii, A. A., 1994. Stabilization and Competitive Inhibition of β-Carotene 15,15′-Dioxygenase by Carotenoids, *Biochem. Mol. Biol. Int.,* 34:755–763.

Fotouhi, N., Meydani, M., Santos, M. S., Meydani, S. N., Hennekens, C. H., and Gaziano, J. M. 1996. Carotenoid and Tocopherol Concentrations in Plasma, Peripheral Blood Mononuclear Cells, and Red Blood Cells after Long-term β-Carotene Supplementation in Men, *Am. J. Clin. Nutr.,* 63:553–558.

Gärtner, C., Stahl, W., and Sies, H., 1996. Preferential Increase in Chylomicron Levels of the Xanthophylls Lutein and Zeaxanthin Compared to β-Carotene in the Human, *Int. J. Vit. Nutr. Res.,* 66:119–125.

Goodman, D. S., and Huang, H. S., 1965. Biosynthesis of Vitamin A with Rat Intestinal Enzymes, *Science,* 149:879–880.

Goodwin, T. W., 1984. *The Biochemistry of the Carotenoids, vol. II,* 2nd ed. New York, NY: Chapman and Hall, pp. 173–195.

Greenberg, S. M., Calbert, C. E., Pinckard, J. H., Deuel, H. J., Jr., and Zechmeister, L., 1949. Stereochemical Configuration and Provitamin A Activity. IX. A Comparison of All-*trans*-γ-Carotene and Pro-γ-Carotene with All-*trans*-β-Carotene in the Chick, *Arch. Biochem.,* 24:31–39.

Gugger, E. T., and Erdman, J. W., 1996. Intracellular β-Carotene Transport in Bovine Liver and Intestine is Not Mediated by Cytosolic Proteins, *J. Nutr.,* 126:1470–1474.

Hébuterne, X., Wang, X.-D., Smith, D. E. H., Tang, G., and Russell, R. M., 1996. *In Vivo* Biosynthesis of Retinoic Acid from β-Carotene Involves an Excentric Cleavage Pathway in Ferret Intestine, *J. Lipid Res.,* 37:484–492.

Hennekens, C. H., Buring, J. E., Manson, J. E., et al., 1996. Lack of Effect of Long-term Supplementation with β-Carotene on the Incidence of Malignant Neoplasms and Cardiovascular Disease, *N. Engl. J. Med.,* 334:1145–1149.

Hollander, D., and Ruble, P. E., 1978. β-Carotene Intestinal Absorption: Bile, Fatty Acid, pH, and Flow Rate Effects on Transport, *Am. J. Physiol.,* 235:E686–E691.

Jensen, C. D., Howes, T. W., Spiller, G. A., Pattison, T. S., Whittam, J. H., and Scala, J., 1987. Observations on the Effects of Ingesting *cis-* and *trans*-Beta-Carotene Isomers on Human Serum Concentrations, *Nutr. Rep. Int.,* 35:413–422.

Johnson, E. J., Krinsky, N. I., and Russell, R. M., 1996. Serum Response of All-*trans* β-Carotene and Lycopene in Humans after Ingestion of Individual and Combined Doses of β-Carotene and Lycopene, *FASEB J.,* 10:A239.

Johnson, E. J., and Russell, R. M., 1992. Distribution of Orally Administered β-Carotene among Lipoproteins in Healthy Men, *Am. J. Clin. Nutr.,* 56:128–135.

Karpe, F., Bell, M., Björkegren, J., and Hamsten, A., 1995. Quantification of Postprandial Triglyceride-rich Lipoproteins in Healthy Men by Retinyl Ester Labeling and Simultaneous Measurement of Apolipoproteins B-48 and B-100, *Arterioscler. Thromb. Vasc. Biol.,* 15:199–207.

Karpe, F., Steiner, G., Olivecrona, T., Carlson, L. A., and Hamsten, A., 1993. Metabolism of Triglyceride-rich Lipoproteins during Alimentary Lipemia, *J. Clin. Invest.,* 91:748–758.

Kemmerer, A. R., and Fraps, G. S., 1945. The Vitamin A Activity of neo-β-Carotene U and Its Steric Rearrangement in the Digestive Tract of Rats, *J. Biol. Chem.,* 161:305–309.

Kostic, D., White, W. S., and Olson, J. A., 1995. Intestinal Absorption, Serum Clearance, and Interactions between Lutein and β-Carotene when Administered to Human Adults in Separate or Combined Doses, *Am. J. Clin. Nutr.,* 62:604–610.

Leo, M. A., and Lieber, C. S., 1994. Beta Carotene, Vitamin E, and Lung Cancer, *N. Eng. J. Med.,* 330:612.

MacDonald, P. N., and Ong, D. E., 1987. Binding Specificities of Cellular Retinol-Binding Protein and Cellular Retinol-Binding Protein, Type II, *J. Biol. Chem.,* 262:10550–10556.

Micozzi, M. S., Brown, E. D., Edwards, B. K., Bieri, J. G., Taylor, P. R., Khachik, F., Beecher, G. R., and Smith, J. C., 1992. Plasma Carotenoid Response to Chronic Intake of Selected Foods and β-Carotene Supplements in Men, *Am. J. Clin. Nutr.,* 55:1120–1125.

Nutting, D., Hall, J., Barrowman, J. A., and Tso, P., 1989. Further Studies on the Mechanism of Inhibition of Intestinal Chylomicron Transport by Pluronic L-81, *Biochim. Biophys. Acta,* 1004:357–362.

Ockner, R. K., and Manning, J., 1974. Fatty Acid-Binding Protein in Small Intestine: Identification, Isolation, and Evidence for its Role in Cellular Fatty Acid Transport, *J. Clin. Invest.,* 54:326–338.

Olson, J. A., 1961. The Conversion of Radioactive β-Carotene into Vitamin A by the Rat Intestine In Vivo, *J. Biol. Chem.,* 236:349–356.

Olson, J. A., and Hayaishi, O., 1965. The Enzymatic Cleavage of β-Carotene into Vitamin A by Soluble Enzymes of Rat Liver and Intestine, *Proc. Natl. Acad. Sci. USA,* 54:1364–1370.

Omenn, G. S., Goodman, G. E., Thornquist, M. D., Balmes, J., Cullen, M. R., Glass, A., Keogh, J. P., Meyskens, F. L., Valanis, B., Williams, J. H., Barnhart, S., Cherniack, M. G., Brodkin, C. A., and Hammar, S., 1996a. Risk Factors for Lung Cancer and for Intervention Effects in CARET, the Beta-Carotene and Retinol Efficacy Trial, *J. Natl. Cancer Inst.,* 88:1550–1559.

Omenn, G. S., Goodman, G. E., Thornquist, M. D., Balmes, J., Cullen, M. R., Glass, A., Keogh, J. P., Meyskens, F. L., Valanis, B., Williams, J. H., Barnhart, S., and Hammar, S., 1996b. Effects of a Combination of Beta Carotene and Vitamin A on Lung Cancer and Cardiovascular Disease, *N. Engl. J. Med.,* 334:1150–1155.

Ong, D. E., 1993. Retinoid Metabolism During Intestinal Absorption, *J. Nutr.,* 123:351–355.

Ong, D. E., and Page, D. L., 1987. Cellular Retinol-Binding Protein (Type Two) Is Present in Human Small Intestine, *J. Lipid Res.,* 28:739–745.

Paetau, I., Chen, H., Goh, N. M. -Y., and White, W. S. In Press. Interactions of the Postprandial Appearance of β-Carotene and Canthaxanthin in Plasma Triacylglycerol-rich Lipoproteins in Humans, *Am. J. Clin. Nutr.* 66:1133–1143, 1997.

Parker, R. S., 1996. Absorption, Metabolism, and Transport of Carotenoids, *FASEB J.*, 10:542–551.

Schneeman, B. O., Kotite, L., Todd, K. M., and Havel, R. J., 1993. Relationships between the Responses of Triglyceride-rich Lipoproteins in Blood Plasma Containing Apolipoproteins B- 48 and B-100 to a Fat-Containing Meal in Normolipidemic Humans, *Proc. Natl. Acad. Sci. USA*, 90:2069–2073.

Scita, G., Aponte, G. W., and Wolf, G., 1992. Uptake and Cleavage of β-Carotene by Cultures of Rat Small Intestinal Cells and Human Lung Fibroblasts, *J. Nutr. Biochem.*, 3:118–123.

Shiau, Y. F., and Levine, G. M., 1980. pH Dependence of Micellar Diffusion and Dissociation, *Am. J. Physiol.*, 239:G177–G182.

Stahl, W., Schwarz, W., and Sies, H., 1993. Human Serum Concentrations of all-*trans*-β- and α-Carotene but not 9-*cis* β-Carotene Increase upon Ingestion of a Natural Isomer Mixture Obtained from *Dunaliella salina* (Betatene), *J. Nutr.*, 123:847–851.

Stahl, W., Schwarz, W., von Laar, J., and Sies, H., 1995. all-*trans* β-Carotene Preferentially Accumulates in Human Chylomicrons and Very Low Density Lipoproteins Compared with the 9-*cis* Geometrical Isomer, *J. Nutr.*, 125:2128–2133.

Tang, G., Serfaty-Lacrosniere, C., Camilo, M. E., and Russell, R. M., 1996. Gastric Acidity Influences the Blood Response to a β-Carotene Dose in Humans, *Am. J. Clin. Nutr.*, 64:622–626.

Thomson, A. B. R., Schoeller, C., Keelan, M., Smith, L., and Clandinin, M. T., 1993. Lipid Absorption: Passing Through the Unstirred Layers, Brush-Border Membrane, and Beyond, *Can. J. Physiol. Pharmacol.*, 71:531–555.

van Vliet, T., van Schaik, F., Schreurs, W. H. P., and van den Berg, H., 1996. In Vitro Measurement of β-Carotene Cleavage Activity: Methodological Considerations and the Effect of Other Carotenoids on β-Carotene Cleavage, *Int. J. Vit. Nutr. Res.*, 66:77–85.

Wahlqvist, M. L., Wattanapenpaiboon, N., Macrae, F. A., Lambert, J. R., MacLennan, R., Hsu-Hage, B. H.-H., and Australian Polyp Prevention Project Investigators., 1994. Changes in Serum Carotenoids in Subjects with Colorectal Adenomas after 24 Mo of β-Carotene Supplementation, *Am. J. Clin. Nutr.*, 60:936–943.

White, W. S., Stacewicz-Sapuntzakis, M., Erdman, J. W., and Bowen, P., 1994. Pharmacokinetics of β-Carotene and Canthaxanthin after Ingestion of Individual and Combined Doses by Human Subjects, *J. Am. Coll. Nutr.*, 13:665–671.

Wirtz, K. W. A., and Gadella, T. W. J., 1990. Properties and Modes of Action of Specific and Nonspecific Phospholipid Transfer Proteins, *Experientia*, 46:592–598.

Wolf, G., 1995. The Enzymatic Cleavage of β-Carotene: Still Controversial, *Nutr. Rev.*, 53:134–137.

You, C.-S., Parker, R. S., Goodman, K. J., Swanson, J. E., and Corso, T. N., 1996. Evidence of *cis-trans* Isomerization of 9-*cis*-β-Carotene during Absorption in Humans, *Am. J. Clin. Nutr.*, 64:177–183.

Zechmeister, L., 1962. *Cis-trans* Isomerism and Provitamin A Effect of Carotenoids, In: Cis-Trans *Isomeric Carotenoids Vitamins A and Arylpolyenes*. New York, NY: Academic Press, pp. 118–145.

CHAPTER 7

Case Study: *Dunaliella* Algal-Derived β-Carotene

AMI BEN-AMOTZ

BIOLOGY AND HALOTOLERANCE

THE biflagellated alga *Dunaliella* is classified under Chlorophyceae, Volvocales, which includes a variety of ill-defined marine and freshwater unicellular species (Avron and Ben-Amotz, 1992). *Dunaliella*, like *Chlamydomonas*, is characterized by an ovoid cell volume usually in the shape of a pear, wider at the basal side and narrow at the anterior flagella top (Figure 1). The cellular organization of *Dunaliella* is no different from that of other members of the Volvocales, presenting one large chloroplast with single-centered starch surrounded by pyrenoid, a few vacuoles, a nucleus, and a nucleolus. However, unlike other green algae, *Dunaliella* lacks a rigid polysaccharide cell wall and is enclosed by a thin elastic plasma membrane covered by a mucous surface coat. The lack of a rigid cell wall permits rapid cell volume changes in response to extracellular changes in osmotic pressure. Commonly osmotically treated *Dunaliella* vary from round to eight shaped. Under extreme salt concentrations greater than 4M *Dunaliella* loses their flagella and the surrounding mucous as the cell rounds with a buildup of a thick surrounding wall to form a dehydration-resistant cyst.

Dunaliella occurs in a wide range of marine habitats such as oceans, brine lakes, salt marshes, and saltwater ditches near the sea, predominantly in water bodies containing more than 2M salt and high levels of magnesium. The effect of magnesium on the distribution of *Dunaliella* is not clear, but, in many "bittern" habitats of marine salt producers, *Dunaliella* usually flourishes. The phenomenon of orange-red algal bloom in such marine environments is usually

113

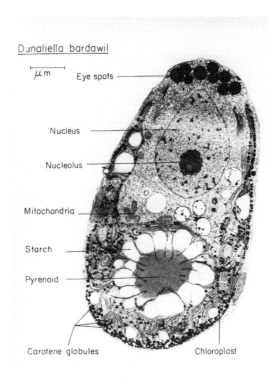

Dunaliella bardawil

Eye spots

Nucleus

Nucleolus

Mitochondria

Starch

Pyrenoid

Carotene globules

Chloroplast

Figure 1 Electron-micrograph of *Dunaliella bardawil,* an ovoid biflagellate, β-carotene-rich halotolerant alga. The dimensions of the alga is in the range of 2–8 μm wide × 5–15 μm in length.

related to combined sequential growth of *Dunaliella* first, and then halophilic bacteria, as commonly observed in concentrated saline lakes such as the Dead Sea in Israel, the Pink Lake in Western Australia, the Great Salt Lake in Utah, and in many other places around the globe. *Dunaliella* is the most halotolerant eukaryotic organism known, showing a remarkable degree of adaptation to a variety of salt concentrations from as low as 0.1M to salt saturation.

Dunaliella uses a special mechanism of osmoregulation to adapt to the various salt concentrations in the surrounding media by varying the intracellular concentration of glycerol in response to the extracellular osmotic pressure. The intracellular concentration of glycerol is directly proportional to the extracellular salt concentration and is sufficient to account for all the required cytoplasmic and most of the chloroplastic osmotic pressures. Glycerol biosynthesis and elimination occur in the light or dark by a novel glycerol cycle that involves several specific enzymes (Avron, 1992).

β-CAROTENE PRODUCTION

Among the many known strains of *Dunaliella,* only a few subspecies of *D. salina* Teod. have been shown to produce and accumulate large amounts of *β*-carotene. In hypersaline lakes, which are generally low in available nitrogen and exposed to high solar radiation, these *β*-carotene-producing strains of *Dunaliella* predominate over all other organisms to a seasonal bloom of about 0.1 mg *β*-carotene per liter. Under such stressful environmental conditions, more than 12% of the algal dry weight is *β*-carotene, usually associated with a sharp decline in the thylakoid chlorophyll. The *β*-carotene in *Dunaliella* accumulates within distinctive oily globules in the interthylakoid spaces of the chloroplast periphery. Analysis of the globules showed that the *β*-carotene of *Dunaliella* is composed mainly of two stereoisomers: all-*trans* and 9-*cis,* with the rest a few other mono-*cis* and di-*cis* stereoisomers of the *β*-carotene (Figure 2). Both the amount of the accumulated *β*-carotene and the 9-*cis* to all-*trans* ratio depend on light intensity and on the algal division time, which is

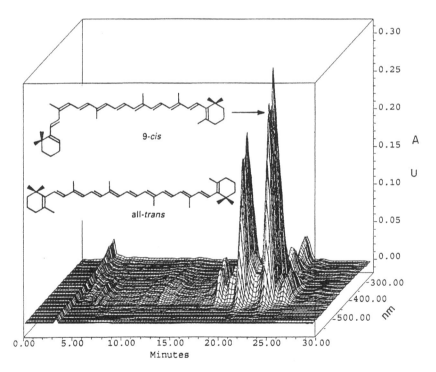

Figure 2 A three dimensional, high pressure liquid chromatogram of pigments extracted from *Dunaliella bardawil.* The predominant peaks are all-*trans* and 9-*cis* *β*-carotene.

determined by the growth conditions. Thus, any growth stress that will slow down the rate of cell division under light will in turn increase β-carotene production in *Dunaliella*. In fact, high light and many environmental stress conditions such as high salt, low temperature, extremes of pH, nutrient deficiencies, and others affect the content of β-carotene in *Dunaliella*. It was previously suggested that the equation of the amount of light absorbed by the cell during one division cycle integrates the effect of all growth variables on the content and isomeric ratio of β-carotene in *Dunaliella* (Ben-Amotz and Avron, 1990). The exceptions to this integration are nitrogen deficiency and low growth temperatures, both of which induce extreme intracellular accumulation of β-carotene under any light intensity. Taking into account that nitrogen starvation inhibits chlorophyll production in algae as part of its inhibitory effect on protein biosynthesis, the prolonged nitrogen-independent biosynthesis of carotenoids will then protect the chlorophyll-reduced cells against the lethal damage of light. The effect of low temperatures on *Dunaliella* can be analyzed by measuring the specific stimulation effect of chilling on the biosynthesis of 9-*cis* β-carotene. The physicochemical properties of 9-*cis* β-carotene differ from those of all-*trans* β-carotene. All-*trans* β-carotene is practically insoluble in oil and is easily crystallized at low temperatures, while 9-*cis* β-carotene is much more soluble in hydrophobic-lipophilic solvents, very difficult to crystallize, and generally oily in its concentrated form. To avoid cellular crystallization of all-*trans* β-carotene and to survive at low temperatures, *Dunaliella* produces a higher ratio of 9-*cis* to all-*trans* β-carotene, where the 9-*cis* stereoisomer functions *in vivo* as an oily matrix for the all-*trans* form (Ben-Amotz, 1996).

CAROTENEOGENESIS

Although 272 geometric isomers of β-carotene can exist theoretically, 12 *cis* forms in total have been noted and recorded. Two *cis* forms, neo-β-carotene B and U, all-*trans* and 9-mono-*cis*, respectively, were reported by Zechmeister (1962). Physicochemical methods for stereomutation of β-carotene involve heat, light with no catalyst, and iodine catalysis under light yielding between one third and one half of the pigment in *cis* configuration. *Dunaliella* follows the same biosynthetic pathway of carotenoids, with the same substrates and same intermediates as those found in other eukaryotic organisms and plants. Mevalonic acid through isopentenyl diphosphate, geranyl diphosphate, and geranylgeranyl diphosphate forms phytoene, which undergoes a few desaturation steps and cyclization to β-carotene (Goodwin, 1988). The isomerization reaction, which produces 9-*cis* β-carotene, is not identified as yet. The observation of two stereoisomers of phytoene, 9-*cis* and all-*trans*, suggests that the biosynthesis of 9-*cis* β-carotene initiates probably early in the pathway of carotene biosynthesis at or before the formation of phytoene.

Thereafter, all the intermediates occur in two isomeric forms. The induction of β-carotene in *Dunaliella* as described above was successfully applied to accumulate any choice of stereoisomeric intermediates in the caroteneogenesis pathway, yielding a large quantity of the twin stereoisomers, all-*trans* and 9-*cis*: phytoene, phytofluene, neurosporene, ξ-carotene, β-zeacarotene, γ-carotene, and β-carotene (Figure 3) (Shaish et al., 1990, 1991). Subsequently, Ebenezer and Pattenden (1993) used ¹H and ¹³C nuclear magnetic resonance

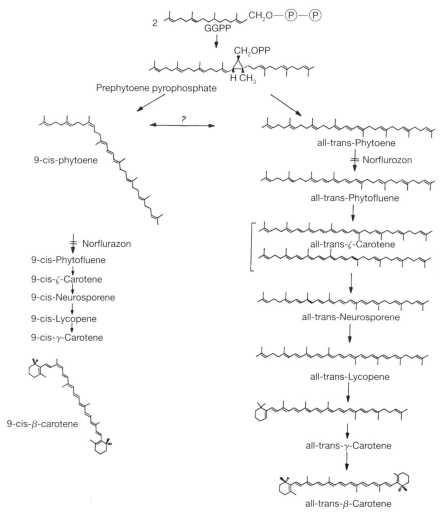

Figure 3 Postulated pathway of all-*trans* and 9-*cis* β-carotene biosynthesis in *Dunaliella bardawil*.

(NMR) on purified phytoene and phytofluene from *D. bardawil* extracts to verify the high-performance liquid chromatography (HPLC) stereoisomeric analysis. Their NMR analysis showed 15-*cis* phytoene and 9-*cis* phytofluene as an indication that the phytoene is the branch point for the formation of 9-*cis* β-carotene. The contradiction between the HPLC and NMR analyses left the intriguing site of stereoisomerization open at any site between isopentenyl diphosphate to phytoene. Furthermore, the induced enzyme(s) or enzyme complex responsible for the high accumulation and isomerization of β-carotene in *Dunaliella* has not been identified as yet.

A few speculative hypotheses have been postulated for the function of the β-carotene globules in *Dunaliella*. One suggests that β-carotene accumulation is related to a protecting effect against chlorophyll-catalyzed singlet oxygen production under high light radiation. However, the electron micrographs of *D. bardawil* (Ben-Amotz and Avron, 1990), which indicate that the massive accumulation of β-carotene occurs in globules located in the interthylakoid space of the chloroplast, are in disagreement with this hypothesis. The large distance between the β-carotene globules and the thylakoid-located chlorophyll would not allow efficient quenching of singlet oxygen or any other chlorophyll-generated product. Other hypotheses concerning the possible function of the globules such as carbon storage for use under limiting growth conditions were eliminated by the observation that β-carotene rich *Dunaliella* cannot utilize or metabolize the β-carotene in the dark nor under carbon dioxide (CO_2)-deficient conditions. The most accepted hypothesis suggests that the β-carotene globules protect the cell against injury by high-intensity radiation under limiting growth conditions by acting as a screen to absorb excess radiation. The harmful effect of the blue region of the spectrum is screened by the algal peripheral-located globules, thus preventing any cellular damage. Strains of *Dunaliella* and other algae unable to accumulate β-carotene die when exposed to high levels of radiation, while the β-carotene-rich *Dunaliella* flourishes. Moreover, protection against photoinhibition by the massively accumulated β-carotene is observed only when the photoinhibitory light is composed of wavelengths absorbed by β-carotene, light in the blue region. No photoprotection at all is observed when red light, which is not absorbed by β-carotene, serves as the photoinhibitory agent. This is in agreement with the observation on the location of the β-carotene globules, distant from the thylakoid-located chlorophyll, and with the above-mentioned hypothesis that the mode of action of the massively accumulated carotene is mostly a screening effect. It was shown previously (Ben-Amotz et al., 1989b) that a series of sequential events leads to the photodestruction of chlorophyll in *D. bardawil*. 9-*cis* β-Carotene is destroyed first, then the all-*trans* β-carotene, and later the chlorophylls, indicating higher sensitivity of 9-*cis* β-carotene to direct elevated levels of light or to damaging chlorophyll-generated free radicals.

The induction of β-carotene biosynthesis in *Dunaliella* is light dependent. Photosynthetic active radiation (PAR) is absorbed by the algal chlorophylls triggering the massive synthesis of β-carotene in *Dunaliella*. The question regarding the involvement of chlorophyll-generated active oxygen species was studied kinetically by adding promoters of oxygen radicals and azide, an inhibitor of catalase and superoxide dismutase, to *Dunaliella* during a selective induction period of β-carotene (Shaish et al., 1993). The actual action of active oxygen species in triggering the biosynthesis of β-carotene remains open for further confirmation.

BIOTECHNOLOGY OF β-CAROTENE PRODUCTION BY *DUNALIELLA*

Dunaliella is a most suitable organism for mass cultivation outdoors in open ponds. The ability to thrive in media with high sodium, magnesium, calcium, and the respective anions, chloride, and sulfate in desert high-solar-irradiated land with access to brackish water or seawater at extreme temperatures from approximately $-5°$ to above $40°C$ make *Dunaliella* most attractive for the biotechnologists and venture capitalists. In fact, since 1980, several firms, government authorities, and industries have invested capital in the application of *Dunaliella* for the production of natural β-carotene. Large-scale *Dunaliella* production is based on autotrophic growth in media containing inorganic nutrients with CO_2 as exclusive carbon sources. Attempts to commercially develop heterotrophic strains or mutants of *Dunaliella* for growth on glucose or acetate, such as *Chlorella* or *Chlamydomonas*, respectively, were not successful. Due to the demand for high light intensity for maximal β-carotene production beyond that required for normal growth, production facilities are located in areas where solar output is maximal and cloudiness is minimal. Most of the present *Dunaliella* production plants are located close to available sources of saltwater, e.g., sea-salt industries usually in warm and sunny areas where the rate of water evaporation is high and nonagricultural land is abundant.

Four modes of cultivation have been used in large-scale production of *Dunaliella*. The first, termed extensive cultivation, uses no mixing and minimal control of the environment. To decrease attacks by zooplanktonic predators, such as certain types of ciliates, amoebase, or brine shrimp, the growers employ very high salt concentrations. *Dunaliella* grows slowly in shallow lagoons in nearly saturated brine, and predators are largely eliminated. The naturally selected strain of *Dunaliella* is well adapted to nearly salt saturation conditions, partially loses its flagella, and produces a thick cell wall on the transformation to a cyst form. Extensive cultivation productivity is low, and the area needed for commercial production is very large, however, the low operating costs of such facilities has led to the development of two commercial plants in Australia and China. The second, termed intensive cultivation, uses high biotechnology to control all factors affecting cell growth and chemistry. The ponds are usually oblong, lined constructed raceways, varying in size up to a production surface area of approximately 3,000 m^2 (Figure 4). The use of long-arm, slow-revolution paddle wheels is presently common in the large-scale facilities in Israel, the United States, China, Chile, and Portugal. One production-sized shallow water pond of 20 cm on an area of 3,000 m^2 (600 m^3) containing 15 g β-carotene per cubic meter yields 9 kg β-carotene on total harvest. The current large-scale production of β-carotene under intensive cultivation is approximately 200 mg β-carotene per square meter per day on a yearly average; thus, a modern intensive plant of 50,000 m^2 produces 1 kg β-carotene per day. Between the extensive and intensive modes, there are examples in Australia and China of a third type, semi-intensive mode, where the ponds are enlarged 10 times, to approximately 50,000 m^2 each with partial control and no mixing. The fourth is highly intensive cultivation in closed photobioreactors. Many trials have been initiated in the last decade to grow *Dunaliella* in different models of closed photobioreactors (Figure 5) with attempts to design the best sunlight-harvesting unit for β-carotene optimization. The different designs include narrow, very long plastic tubes, plastic bags, trays, and more. However, as of today, none of these trials has taken production beyond the laboratory or small pilot-plant volume, mainly due to

Figure 4 A typical commercial cultivation pond of *Dunaliella*. Semi-intensive mode of cultivation. Pond length, 150 m, depth 20 cm, total volume of 600 m^3. Plant location, NBT Ltd., Eilat, Israel.

Figure 5 Pilot large-scale tubular closed bioreactor for cultivation of *Dunaliella*. Highly intensive mode of cultivation. Notice the construction of long clear plastic tube to provide exposure to a daily cycle of 270° solar radiation. Plant location, PBL Ltd., Murcia, Spain.

economic limitations and nonfeasible large-scale development. The few industrial ventures of high intensive-closed photobioreactors became insolvent and no longer exist.

Generally, large-scale optimization of β-carotene production is achieved in all modes of cultivation by high salt stress and by nitrogen deficiency. For pragmatic reasons, the first is applied in the extensive mode, while nitrogen starvation controls the intensive mode. Most species of *Dunaliella* grow optimally in a medium containing 1 to 2 M NaCl in accordance with the medium temperature, exhibiting closely similar growth rates at moderate temperature of 25°C in 2 M NaCl, and at low temperatures of 15°C in 1 M NaCl. The algal composition changes, respectively, by selective accumulation of glycerol and starch. This unique environmental adaptation of *Dunaliella* allows successful intensive outdoor growth in cold seasons and in cold areas such as the Ukraine and Inner Mongolia. Most commercial *Dunaliella* ponds employ evaporated concentrated seawater or seawater augmented with dry salt to reach the desired concentration in the medium. Favored sites for *Dunaliella* cultivation are along the seashore or close to salt lagoons and salt-producing industries for use of a mixture of seawater and concentrated seawater in order to obtain the desired salt concentration by season and temperature. The one exception is the Nutrilite Division of Amway Corporation, located far from the sea in Imperial Valley, CA, which employs freshwater augmented with commercial salt. The use of recycled high-salt medium is common in a few plants after harvesting of the algae by oxidative treatment of the organic load and filtration. The use of recycled medium enriches the medium with higher concentrations of magnesium, calcium, and sulfate. *Dunaliella* was found to grow well in seawater-based media containing approximately 1.5 M NaCl, more than 0.4 M $MgSO_4$, and 0.1 M $CaCl_2$ under strict pH control.

Dunaliella adapts to a broad range of pH, from 5.8 to 11. In natural fresh-water or seawater, several calcium salts, most notably carbonates, sulfates, and phosphates tend to precipitate, causing algal flocculation and a reduction in growth rate. It is, therefore, important to maintain the pH in the *Dunaliella* culture within the neutral range (pH ~ 7) and to avoid an increase of pH above the salting-out range of the divalent cation salts. In autotrophic algal cultures, the pH rises mostly due to photosynthesis. Uncontrolled autotrophic algal cultures show an increase in pH during the day, which can exceed pH 10. Intensive algal growers use pure liquid CO_2 and commercial pH con-trollers to maintain the pH at a preset level. The CO_2 bubbles diffuse through commercial gas diffusers from outlets in the bottom of the pond, dissolve to carbonic acid, lower the pH, and then equilibrate to bicarbonate and carbon-ate as a function of the hydrogen-ion concentration in the medium and finally may be used in photosynthesis. CO_2 consumption in commercial algal plants is higher in the summer and minimal in the winter due to the lower solubility of the CO_2 gas in the medium at elevated temperatures, varying from approxi-mately 6 g CO_2 per gram dry *Dunaliella* to 3 g CO_2 per gram, respectively. Today, large-scale intensive algal cultivation employs modern computerized pH control of many ponds simultaneously by pH sensors in each pond, trans-mitting amplified outputs to a central control room. Supply of CO_2 is moni-tored and controlled continuously by the central pH computer, thus allowing large-scale intensive cultivation. Alternatively, HCl and bicarbonate can be used to maintain the required pH and photosynthesis in the pond. Bicarbonate may avoid the above-mentioned diffusion-solubility and may replace CO_2; however, the market cost of bicarbonate together with HCl and the safety is-sue of the latter are prohibiting factors in industrial operation.

EXTRACTION AND CONCENTRATION

Separation and harvesting of *Dunaliella* out of the growth medium from approximately 0.1% biomass to a concentrated algal paste of more than 15% should take into consideration the specific features of the alga: (1) lack of cell wall and enclosure of the cell by fragile cell membrane, (2) an outer cell layer, termed a surface coat, which possesses mucus properties, (3) elasticity of the cell shape, and (4) high extracellular salinity. Most of the previous ap-proaches to harvest *Dunaliella* by different modes of filtration were not successful and have not reached the level of scale up, suffering mainly from the presence of glutinous surface coat and from the high elasticity of the cell membrane that blocks permeation through many filters and screens. Industrial extensive producers of *Dunaliella* are currently using flocculation and cell surface absorption, while the intensive growers employ centrifugation and flocculation. Flocculation combined with sedimentation or flotation is being used extensively to harvest *Dunaliella.* The unicellular algae flocculate on the addition of inorganic or organic flocculants such as aluminum sulfate (alum), ferrous, ferric chloride, ferric sulfate, lime, polyelectrolytes, guar, polysaccharides, etc. Alum is extensively used to flocculate *Dunaliella* in high-salt media at variable concentrations in relation to pH, temperature, dissolved organic load, bacteria, and mineral concentration. Flocculated *Dunaliella* cannot be used directly for the food market unless the flocculant is safe or is completely released from the algae prior to utilization. Different plants extract the β-carotene from the flocculated *Dunaliella,* leaving behind the aluminum-contaminated cell debris. Centrifugation has been widely used for harvesting *Dunaliella* and other unicellular algae. The most common design is continuous-flow, stainless steel, batch-type centrifuge, or desludge-type centrifuge (self-discharging), operating at large scale at about 1 m³/h, at 15,000 rpm, or 15 m³/h, at 5,000 rpm, respectively. Initial investment is high, and operation energy is costly, but the high efficiency and the clean product attract many industries to employ such centrifuges. The discharged paste varies from 40% algal dry weight in the batch type down to 20% in the desludge type. Another method of harvesting *Dunaliella,* hydrophobic binding, is used by Betatene, Ltd., Australia. *Dunaliella* grown at salinity higher than 4 M converts to cysts and acquires the above-mentioned hydrophobic surface coat. These coated cells adsorb to hydrophobic surfaces and are removed out of the growth suspension to yield algal paste ready for successive β-carotene extraction. The method is most suitable for extensive cultivation where the concentration of *Dunaliella* is below 0.1 g/L and the volume for daily harvest exceeds 20,000 m³.

   β-Carotene can be extracted directly from the wet *Dunaliella* paste, and this is the process employed in several production facilities by different extraction processes using edible oil with or without organic solvents, liquid CO_2 extraction, crystallization, and other common food technology methods. The need for dry *Dunaliella* powder as a product for the health food market has led to the development of several techniques for processing and dehydrating *Dunaliella* paste while protecting its oxygen-sensitive stereoisomers of β-carotene. Several methods have been employed to dry microalgae such as *Chlorella, Scenedesmus,* and *Spirulina* (Ben-Amotz and Avron, 1990), and all are based on conventional food technology for drying different food products such as milk powder, egg powder, tomato powder, baby food powder, and coffee. Most common are spray dryers that work with input of the paste at high temperature (> 150°C) for a short holding time and on output temperature below 100°C, yielding dry algal powder of low moisture content (< 5%). The higher sensitivity of 9-*cis* β-carotene to oxidation and to thermal destruction may lead to rapid loss of total β-carotene in association with a reduction in the 9-*cis* to all-*trans* β-carotene ratio during the drying process. Other techniques of drying, such as drum drying, freeze drying, or sun drying, have been tested but not employed with *Dunaliella* due to the lack of product uniformity and the subsequent need to grind the dry material. Similar problems are encountered on drying different food and feed products, but the high sensitivity of *Dunaliella* and β-carotene to oxidation enforces a specific technological approach.

NATURAL VERSUS SYNTHETIC β-CAROTENE

β-Carotene has been used for many years as a food coloring agent, as provitamin A (retinol) in food and animal feed, as an additive to cosmetics and multivitamin preparations, and in the last decade as a health food product under the "antioxidant" claim. Many epidemiological and oncological studies suggest that humans fed a diet high in carotenoid-rich vegetables and fruits, who maintain higher-than-average levels of serum carotenoids, have a lower incidence of several types of cancer and cardiovascular disease (Krinsky, 1989). Recently, the Alpha-Tocopherol, β-Carotene Cancer Prevention Study Group and the β-Carotene and Retinol Efficacy Trial clearly showed that, not only did β-carotene fail to reduce the incidence of cancer and cardiovascular disease, but in fact it increased it in smokers and workers exposed to asbestos (Omenn et al., 1996; Hennekens et al., 1996). These two U.S. National Cancer Institute-supported studies and numerous earlier trials that dealt with the question of the protective role of β-carotene against chronic diseases used only synthetically formed all-*trans* β-carotene.

Despite the fact that the most convincing reports support a direct connection between high intake of fruits and vegetables and low incidence of cancer and cardiovascular disease, the literature lacks specific information on the possible medical contribution of natural carotenoids, isomers of carotenoids, and carotenoid fatty acid esters. Experimental nutrition and medical studies with natural carotenoids originating from different plants, fruits, vegetables, and algae have been very limited, and such research is in its infancy.

β-Carotene is present in most plants and algae in small amounts of ~0.2% of the dry weight with approximately one third as 9-*cis* β-carotene. The observations of the high β-carotene content of *Dunaliella* containing more than 50% 9-*cis* β-carotene gave impetus to studies on the metabolism, storage, and function of 9-*cis* β-carotene in animals and humans, with emphasis on the possible role of the 9-*cis* stereoisomer in scavenging reactive oxygen species. The first dietary studies showed that low doses of the algal β-carotene are as potent as doses of synthetic all-*trans* β-carotene in providing retinol in rats and chicks (Ben-Amotz et al., 1989a). Later, preferential and selective uptake of either all-*trans* or 9-*cis* β-carotene was noted in different animals and humans. 9-*cis* β-Carotene is not detected in serum of chicks and rats nor in humans fed a diet rich in *Dunaliella* (Ben-Amotz and Levy, 1996) or oil extract of *Dunaliella* (Stahl et al., 1993, Gaziano et al., 1995). The β-carotene detected in any of these sera was all-*trans* β-carotene. Rats and chicks that generally do not accumulate β-carotene in their tissues showed uptake of β-carotene into the tissues when the diet was supplemented with β-carotene-rich *Dunaliella*. The lack of 9-*cis* β-carotene in mammals serum and the different response to central nervous system toxicity (Bitterman and Ben-Amotz, 1994) and to *in vitro* oxidation (Levin and Mokady, 1994) led to the hypothesis that the isomeric structure of 9-*cis* β-carotene acts as a quencher of singlet oxygen and other free radicals serving as effective *in vivo* antioxidant. Generally, carotenoids exert an antioxidative effect by a mechanism that results in the formation of new products that are more stable and more polar. Different carotenoids exhibit different antioxidative-antiperoxidative capacities (Krinsky, 1989). Opening of the β-ionone ring, the addition of a chemical group on the ring, or replacement of the ring by various groups can modify the antioxidative activity. Therefore, structural variables other than the length of the polyene chain may direct the scavenging properties of the carotenoids. Insertion of any *cis* position along the conjugated double-bond chain may lead to modified antioxidative properties. Assuming that the *cis* conformational change leads to a higher steric interference between the two parts of the carotene molecule, the *cis* polyenic chain will be less stable and more susceptible to low oxygen tension in *in vivo* oxidation. Recent studies strongly supported this hypothesis by measuring different antioxidative activity (Levy et al., 1995, 1996) and much lower concentration of low-ultraviolet dienes in the serum of humans supplemented with β-carotene-rich *Dunaliella* compared with all-*trans* β-carotene (Ben-Amotz and Levy, 1996). However, the possibilities of low intestinal absorption, isomerization, or tissue uptake have to be considered as well (Johnson et al., 1996). The possibility of synergistic effects and the possible beneficial potency of the different plant nutrients and carotenoids are still obscure and warrant further research.

DUNALIELLA MARKET PRODUCTS

Dunaliella natural β-carotene is widely distributed today in many different markets under two different categories, β-carotene extracts and dry *Dunaliella*. Extracted purified β-carotene is sold mostly in vegetable oil in bulk concentrations from 1 to 20% or for personal use in soft gels, 5 mg β-carotene each gel. Due to the lack of fine purification during extraction, the β-carotene is generally accompanied by the other carotenoids of *Dunaliella*, predominantly lutein, neoxanthin, zeaxanthin, violaxanthin, cryptoxanthin, and α-carotene, each with its different stereoisomers, comprising approximately 15% of the β-carotene concentration. Under the same category of extracts, one can find granular powder of microencapsulated β-carotene (2–7%) and crystals of β-carotene packed under vacuum or nitrogen atmosphere. A variety of such formulations of β-carotene are currently found and distributed in the world under the market sections "health food" or "food supplement."

The second category covers a line of dried *Dunaliella* powders, intact algae, harvested and dried as described above for marketing in the form of tablets or hard capsules containing between 3 and 20 mg β-carotene per unit. To avoid β-carotene oxidation, the tablets are coated with sugar and the capsules are packed separately in aluminum/polyethylene blisters. Dry *Dunaliella* is distributed popularly in Japan and in other Far East countries where the consumers are more familiar with edible algae and accustomed to *Chlorella* and *Spirulina.*

COMMERCIAL PRODUCERS

At present, the following companies are actively engaged in cultivating *Dunaliella* for commercial purposes. These are and their mode of cultivation is

(1) Betatene Ltd., Cheltenham, Victoria 3192, Australia, a division of Henkel Co., Germany; extensive mode

(2) Cyanotech Corp., Kailua-Kona, HI 96740, USA; intensive mode

(3) Inner Mongolia Biological Eng. Co., Inner Mongolia, 750333, P. R. China; semi-intensive mode

(4) Nature Beta Technologies (NBT) Ltd., Eilat 88106, Israel, a subsidiary of Nikken Sohonsha Co., Gifu, Japan; intensive mode

(5) Nutrilite, Calipatria, CA 92233, USA; a division of Amway Corporation, USA; intensive mode

(6) Photo Bioreactors, PBL Ltd., Murcia 30003, Spain; highly intensive mode, closed photobioreactors

(7) Tianjin Lantai Biotechnology, Inc. Nankai, Tianjin, in collaboration with the Salt Scientific Research Institute of Light Industry Ministry, P. R. China; intensive mode

(8) Western Biotechnology Ltd. Bayswater, W. A. 6053, Australia, a subsidiary of Coogee Chemicals Pty. Ltd., Australia; semi-intensive mode

REFERENCES

Avron, M., 1992, Osmoregulation, in Avron, M. and Ben-Amotz, A. (Eds.) Dunaliella: *Physiology, Biochemistry, and Biotechnology,* pp. 135–164, Boca Raton: CRC Press.

Avron, M, and Ben-Amotz, A. (Eds.), 1992, Dunaliella: *Physiology, Biochemistry, and Biotechnology,* Boca Raton: CRC Press.

Ben-Amotz, A., 1996, Effect of low temperature on the stereoisomer composition of β-carotene in the halotolerant alga *Dunaliella bardawil* (Chlorophyta) *J. Phycol.,* 32:272–275.

Ben-Amotz, A., and Avron, M., 1990, The biotechnology of cultivating the halotolerant alga *Dunaliella, Trends Biotechnol.,* 8:121–26.

Ben-Amotz, A., and Levy, I., 1996, Bioavailability of a natural isomer mixture compared with synthetic all-*trans* β-carotene in human serum, *Am. J. Clin. Nutr.,* 63:729–734.

Ben-Amotz, A., Mokady, S., Edelstein, S., and Avron, M., 1989a, Bioavailability of natural isomer mixture as compared with synthetic all-*trans* β-carotenein rats and chicks, *J. Nutr.,* 119:1013–1019.

Ben-Amotz, A., Shaish, A., and Avron, M., 1989b, Mode of action of the massively accumulated β-carotene of *Dunaliella bardawil* in protecting the alga against damage by excess radiation. *Plant Physiol.,* 91:1040–1043.

Bitterman, N., and Ben-Amotz, A., 1994, β-Carotene and CNS toxicity in rats, *J. Appl. Physiol.,* 76:1073–1076.

Ebenezer, W. J., and Pattenden, G., 1993, *cis*-Stereoisomers and β-carotene and its congeners in the alga *Dunaliella bardawil,* and their biogenetic interrelationships. *J. Chem. Soc. Perkin Trans.,* 1:1869–1873.

Gaziano, J. M., Johnson, E. J., Russel, E. M., et al., 1995, Discrimination in absorbance or transport of β-carotene isomers after oral supplementation with either all-*trans* or 9-*cis* β-carotene, *Am. J. Clin. Nutr.,* 61:1248–1252.

Goodwin, T. W. (Ed.) 1988, *Plant Pigments.* London: Academic Press.

Hennekens, C. H., Buring, J. E., Manson, J. E., Stampfer, M., Rosner, B., Cook, N. R., Belanger, C., LaMotte, F., Gaziano, J. M., Ridker, P. M., Willett, W., and Peto, R. 1996. Lack of effect of long-term supplementation with beta-carotene on the incidence of malignant neoplasms and cardiovascular disease. *N. Engl. J. Med.,* 334:1145–1149.

Johnson, J. E., Krinsky, N. I., and Russell, R. M., 1996, Serum response of all-*trans* and 9-*cis* isomers of β-carotene in humans, *Circulation,* 15:620–624.

Krinsky, N. I., 1989, Antioxidant functions of carotenoids, *Free Radic. Bio. Med.,* 7:617–635.

Levin, G., and Mokady, S., 1994, Antioxidant activity of 9-*cis* compared to all-*trans* β-carotene *in vitro, Free Radic. Biol. Med.,* 17:77–82.

Levy, Y., Ben-Amotz, A., and Aviram, M., 1995, Effect of dietary supplementation of different β-carotene isomers on lipoprotein oxidative modification, *J. Nutr. Environm. Med.,* 5:13–22.

Levy, Y., Kaplan, M., Ben-Amotz, A., and Aviram, M., 1996, Effect of dietary supplementation of β-carotene on human monocyte-macrophage-mediated oxidation of low density lipoprotein. *Isr. J. Med. Sci.* 32:473–478.

Omenn, G. S., Goodman, G. E., Thornquist, M. D., Balmes, J., Cullen, M. R., Glass, A., Keogh, J. P., Meyskens, F. L., Valanis, B., Williams, J. H., Barnhart, J., and Hammar, S., 1996, Effects of combination of beta-carotene and vitamin A on lung cancer and cardiovascular disease, *N. Engl. J. Med.,* 334:1150–1155.

Shaish, A., Avron, M., and Ben-Amotz, A., 1990, Effect of inhibitors on the formation of stereoisomers in the biosynthesis of β-carotene in *Dunaliella bardawil, Plant Cell Physiol.,* 31:689–696.

Shaish, A., Avron, M., Pick, U., and Ben-Amotz, A., 1993, Are active oxygen species involved in induction of β-carotene in *Dunaliella bardawil? Planta,* 190:363–368.

Shaish, A., Ben-Amotz, A., and Avron, M., 1991, Production and selection of high β-carotene mutants of *Dunaliella bardawil* (Chlorophyta), *J. Phycol.,* 27: 652–656.

Stahl, W., Schwartz, W., and Sies, H., 1993, Human serum concentrations of all-*trans* and α-carotene but not 9-*cis* β-carotene increase upon ingestion of a natural mixture obtained from *Dunaliella salina* (Betatene), *J. Nutr.,* 123:847–851.

Zechmeister, L., 1962, cis-trans *Isomeric Carotenoids, Vitamin A, and Arylpolyenes,* Vienna: Spriger-Verlag.

CHAPTER 8

The Organosulfur and Organoselenium Components of Garlic and Onions

ERIC BLOCK

INTRODUCTION

GARLIC (*Allium sativum*) and onion (*A. cepa*), as well as related genus *Allium* plants such as leek, shallot, chive, and scallion, respectively (*A. porrum* L., *A. ascalonicum* auct., *A. schoenoprasum* L., and *A. fistulosum* L.), are immensely popular, economically important spices and vegetables. The antibiotic, anticancer, antithrombotic, cholesterol-lowering, and other beneficial health effects associated with consumption of these plants are widely touted in the popular and scientific/medical press.[1,2] Typical culinary usage involves cutting or crushing the plants to maximize flavor and aroma release. Cutting or crushing results in disruption of plant tissue with ensuing enzymatic and chemical reactions generating the aroma and flavor compounds.[3] The health benefits associated with consuming *Allium* species may be attributed to compounds found in the intact plants, flavorants formed on cutting or crushing the plants, substances derived from further reactions of these flavorants, or metabolic degradation products of these 3 types of compounds.[1-3] Most of the *Allium* species compounds of interest from either a health or flavor standpoint contain sulfur, often in forms rarely found elsewhere in nature. The focus of our research, and the subject of this chapter, is the identification and chemical study of such compounds, along with related *Allium* spp. compounds containing the element selenium, with the overall goal of better understanding both the flavor chemistry and the medicinal chemistry of genus *Allium* plants.

129

THIOSULFINATES IN *ALLIUM* HOMOGENATES: GARLIC ORGANOSULFUR COMPOUNDS

In 1945, Chester Cavallito and coworkers in Rennselaer, New York, reported that, when garlic cloves were frozen in dry ice, pulverized and extracted with acetone,

The acetone extracts upon evaporation yielded only minute quantities of residue and no sulfides, indicating the absence of free sulfides in the plant. . . . The [white garlic] powder had practically no odor, but upon addition of small quantities of water, the typical odor was detected and the antibacterial principle could be extracted and isolated. This demonstrates that neither . . . [the antibacterial principle] nor the allyl sulfides found in "Essential Oil of Garlic" are present as such in whole garlic. . . When the powder was heated to reflux for thirty minutes with a small volume of 95% ethanol, no activity could be demonstrated by addition of water to the insoluble residue. . . . When, however, a small quantity (1 mg per cc) of fresh garlic powder was added to the alcohol insoluble fraction in water (20 mg per cc), the activity of the treated sample was shown to be equal to that of the original untreated powder. . . . The 95% ethanol treatment has inactivated the enzyme required for cleavage of the precursor and addition of a small quantity of fresh enzyme brought about the usual cleavage.[4] (p. 1032)

The stable precursor of Cavallito's antibacterial principle of garlic is (+)-*S*-2-propenyl-L-cysteine *S*-oxide (**1a**, "alliin").[3] In intact garlic cloves **1a** and related *S*-alk(en)yl-L-cystein *S*-oxides (aroma and flavor precursors) are physically separated from the C-S lyase enzyme alliinase, with the latter being stored in the relatively few vascular bundle sheath cells located around the veins and the former being concentrated in the abundant storage mesophyll cells.[1] Disruption of the cells results in release of alliinase and subsequent C-S bond cleavage of sulfoxides such as **1a,** ultimately affording volatile and odorous low molecular weight organosulfur compounds such as *S*-2-propenyl 2-propenethiosulfinate ($CH_2 = CHCH_2S(O)SCH_2CH = CH_2$, **3a**, allicin) by a process thought to involve the intermediacy of the sulfenic acid 2-propenesulfenic acid (**2a**). Four sulfoxides occur in *Allium* spp.: *S*-2-propenyl-, *S*-(*E*)-1-propenyl-, *S*-methyl-, and *S*-*n*-propyl-L-cysteine *S*-oxides (**1a–d**, respectively). Garlic contains **1a** to **c**, with only a trace of **1d**, while onions contain **1b** to **d**, with only a trace of **1a**. Alliin is produced in garlic (or in the laboratory) by oxidation of *S*-2-propenyl-L-cysteine. Allicin (**3a**) and related thiosulfinates can be synthesized by oxidation of the corresponding disulfides, e.g., diallyl disulfide (**4**, Scheme 1).

Chromatography of a solution prepared by soaking chopped garlic in methanol affords allicin (**3a**), diallyl di-, tri-, and tetrasulfide (**4a–c**), allyl methyl trisulfide, 3-vinyl-4*H*-1,2-dithiin (**5**), and 2-vinyl-4*H*-1,3-dithiin (**6**), along with two isomeric polar compounds, termed (*E*)- and (*Z*)-ajoene (based

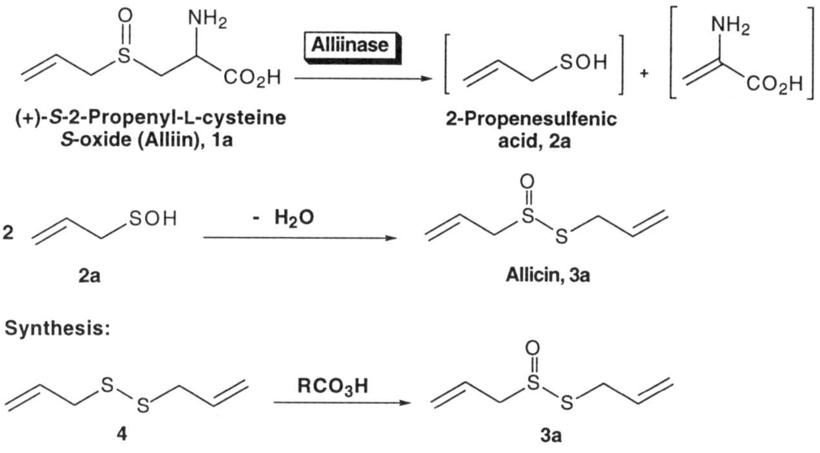

(+)-S-2-Propenyl-L-cysteine S-oxide (Alliin), 1a → **Alliinase** → [**2-Propenesulfenic acid, 2a**] + [NH₂ / CO₂H]

2 (2a) — $- H_2O$ → **Allicin, 3a**

Synthesis:

4 — RCO_3H → 3a

Scheme 1

on the Spanish word for garlic, ajo, pronounced "aho"), characterized as (*E,Z*)-4,5,9-trithiadodeca-1,6,11-triene 9-oxide (**7**) by spectroscopic and synthetic methods. Ajoene is presumably formed as shown in Scheme 2.[5] On injection into a gas chromatograph (GC), allicin (**3a**) decomposes into thioacrolein (**8**) Diels-Alder adducts **5** and **6**.[5] It is of interest that garlic extract components **5, 6**, and **7** all show antithrombotic activity.[5]

A variety of polysulfides has been detected by GC and GC-mass spectrometry (MS) in extracts, volatiles, and distilled oils of garlic and other *Allium* species. In some cases, these compounds are artifacts resulting from decomposition of thiosulfinates such as allicin during the isolation or in the GC injection port.[6,7] In our work, we minimize problems due to artifacts by (1) employing the particularly mild and rapid technique of supercritical fluid extraction (SFE) using liquified carbon dioxide for flavorant extraction and isolation,[8,9] (2) using cryogenic GC-MS for analysis of thiosulfinates,[10] other than those with allyl groups, in parallel with the room temperature techniques of high-performance liquid chromatography (HPLC)[11] or liquid chromatography-atmospheric pressure chemical ionization-MS (LC-APCI-MS),[12] and (3) synthesizing all thiosulfinates, and related compounds, for unambiguous differentiation among regio- and stereoisomers and to validate the analytical methods.[13,14] Analysis of SF extracts of garlic by these methods indicated the presence of allicin (>50% of total thiosulfinates), allyl/methyl and 1-propenyl/methyl isomers (>25%), allyl/1-propenyl isomers (>19%), traces of allyl/*n*-propyl isomers, and ajoene.[13]

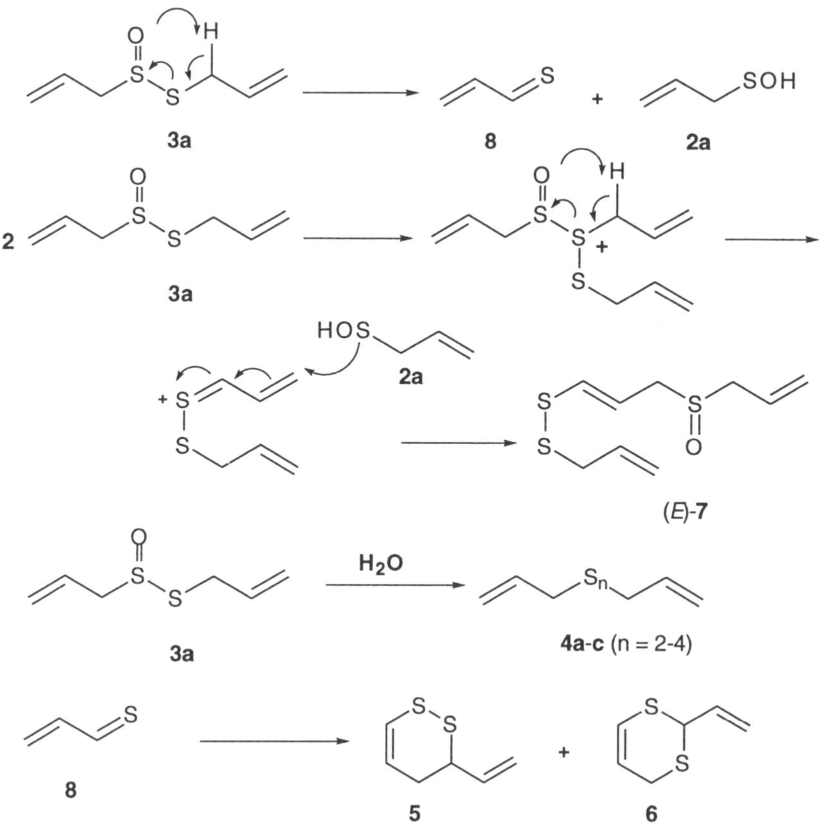

Scheme 2

THE LACHRYMATORY FACTOR (LF) AND OTHER ONION ORGANOSULFUR COMPOUNDS

The ability of the onion to bring tears to those that would cut it has been widely recognized since earliest times. Thus, Shakespeare writes, "Indeed the tears live in an onion that should water this sorrow."[15] The major cysteine S-oxide present in an intact onion is S-(E)-1-propenyl-L-cysteine S-oxide (**1b**; isoalliin). When an onion is cut, **1b** is rapidly cleaved by an alliinase enzyme, giving (E)-1-propenesulfenic acid (MeCH = CHS-O-H; **2b**), which re-arranges to (Z)-propanethial S-oxide (EtCH = S + $-O^-$; **9**), the onion LF.[16] The mechanism for LF formation shown in Scheme 3 is supported by the ob-servation that, from onions macerated in the presence of D_2O, (Z)-**9**-d_1 is de-

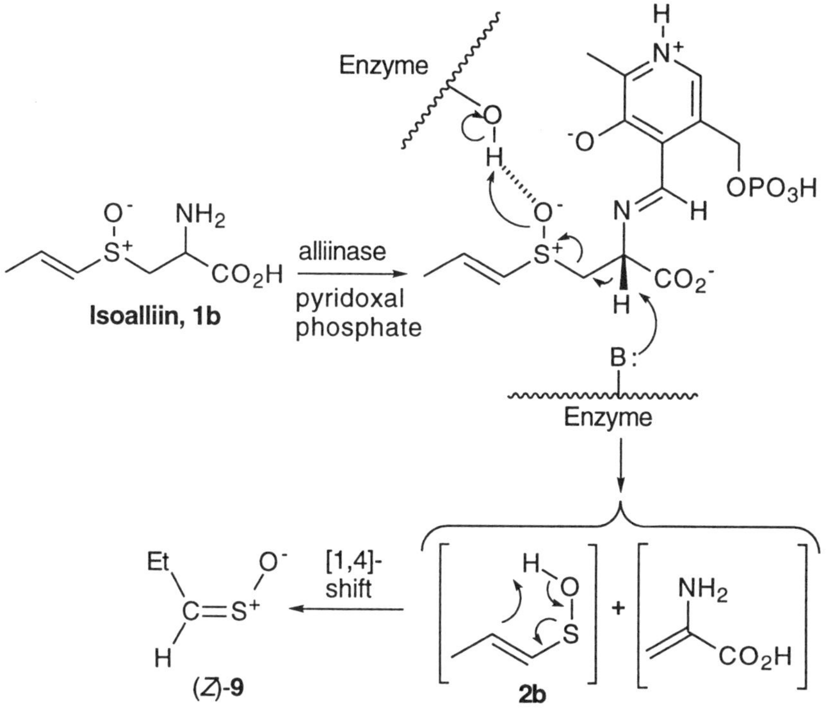

Scheme 3

tected by Fourier transform-microwave (FT-MW) spectroscopy. Compounds (Z)-9-d$_1$ and (Z)-9 are also detected by FT-MW spectroscopy when *tert*-butyl (E)-1-propenyl sulfoxide (**10b**) is subjected to flow pyrolysis (400°C, 0.01 mm) in the presence of D$_2$O (Scheme 4).[16] The LF can be easily quantified by GC.[17] Sulfenic acid **2b** also undergoes condensation with sulfenic acids such as methanesulfenic acid (**2c**), forming thiosulfinates MeCH = CHS(O)SMe/MeCH = CHSS(O)Me (**3b**).

Additional mechanistic support comes from work on methanesulfenic acid (**2**), generated in the gas phase by flash vacuum pyrolysis of **tert**-butyl methyl sulfoxide (**10**a) (Scheme 5). By microwave spectroscopy, a useful technique for detection and characterization of reactive small molecules, **2c** is found to have the structure MeS-O-H [rather than MeS(O)H] and a gas phase life time of approximately 1 minute at 0.1 Torr and 25°C. When condensed at −196°C, **2c** could not be recovered on warming in vacuum, instead losing water to form thiosulfinate MeS(O)SMe (**3c**). In the gas phase in the presence of D$_2$O, MeSOH is readily converted into MeSOD.[18]

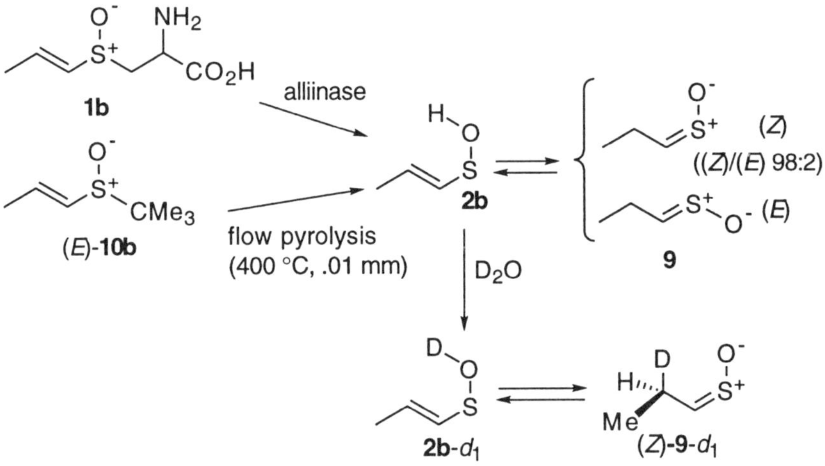

Scheme 4

An impressive variety of acyclic, cyclic, and bicyclic compounds of novel structure can be detected and isolated in extracts of homogenized onions. Among these compounds are the bissulfine **11**, zwiebelanes **12a,b** (from the German word for onion, *Zwiebel*), cepaene **13** (from the botanical name for onion, *Allium cepa*), and LF dimer **14** (Scheme 6). All of these compounds can be independently synthesized, for example, in the case of **11** and **12**, by mono- or dioxidation of bis-1-propenyl disulfide **15**, respectively. The zwiebelanes are thought to be formed via a [3,3]-sigmatropic rearrangement followed by a [2 + 2]-cycloaddition process, as shown in Scheme 6. With the use of the above-described carbon dioxide SFE methods coupled with LC-

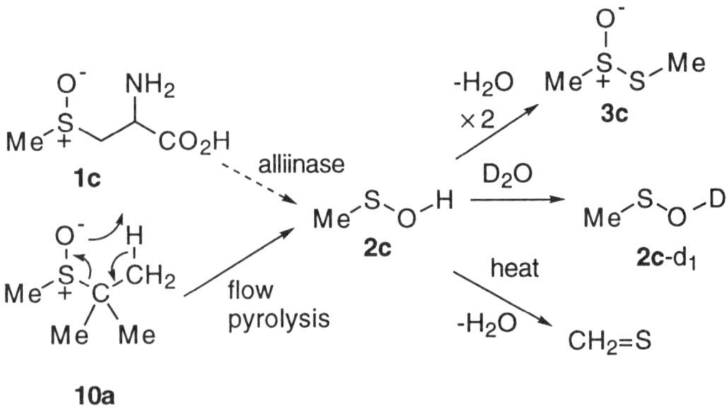

Scheme 5

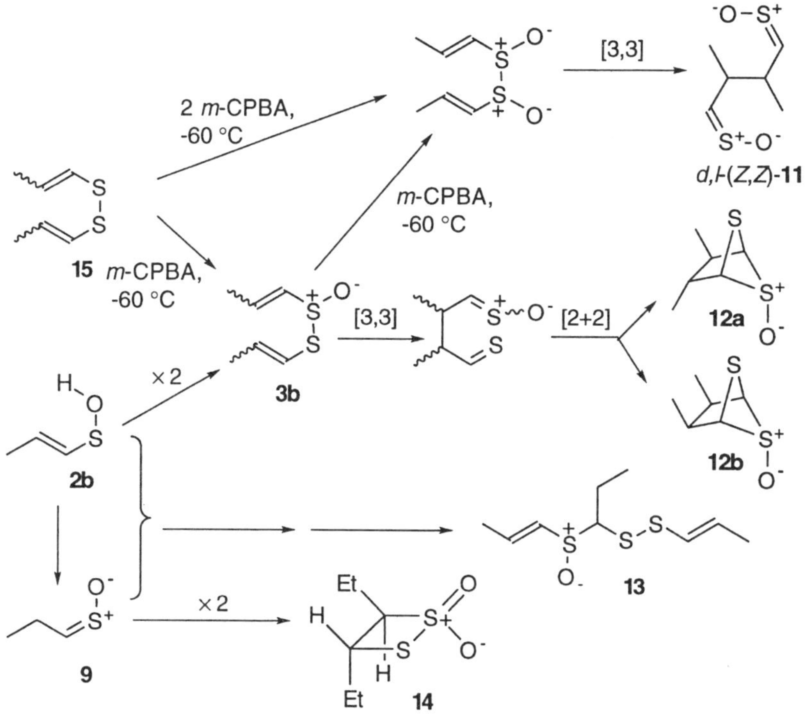

Scheme 6

APCI-MS/MS analysis, with synthetic standards for calibration purposes, more than 25 individual sulfur compounds could be positively identified in onion homogenates.[12] These included methyl/methyl, methyl/1-propenyl, methyl/propyl, 1-propenyl/propyl, propyl/propyl, and traces of 2-propenyl/1-propenyl thiosulfinates in addition to LF **9**, compounds **11** to **13**, and homologues.[12]

ORGANOSELENIUM COMPOUNDS IN GARLIC AND ONION

In 1964, Finnish Nobel Laureate A. I. Virtanen reported on the basis of radioisotope studies that the selenoamino acids selenocystine [HOOCCH(NH$_2$)CH$_2$Se)$_2$] and selenomethionine [HOOCCH(NH$_2$)CH$_2$CH$_2$SeMe] were present in onion.[19] Virtanen's results suggested that there might be a selenium (Se)-based flavor chemistry in *Allium* spp. parallel to that based on sulfur, namely originating from soil selenate (SeO_4^{-2}) or selenite (SeO_3^{-2}) rather than sulfate. While the natural abundance of sulfur in garlic and onion is typically more than 10,000 times higher than that of Se, Se enrichment could occur in crops grown in regions containing higher-than-average-levels of Se in the soil, such as parts of central California. Identification of natural organoselenium compounds is of considerable current interest in view of the discovery that Se-enriched garlic[20] as well as yeast[21] possess cancer-preventative properties. We have obtained information on the nature of organoselenium compounds in garlic and related *Allium* species as well as information on how the body handles the consumed Se compounds.

Detection of natural Se compounds, admixed with far higher levels of chemically similar sulfur compounds, requires the use of element-specific analytical methods such as GC-atomic emission detection (GC-AED), high performance ion chromatography-inductively coupled plasma-MS (HPIC-ICP-MS), and HPLC-ICP-MS. Using the technique of GC-AED, the Se emission line at 196 nm is monitored to identify organoselenium species while concurrently monitoring S and C by lines at 181 and 193 nm, respectively; assignments were confirmed by GC-MS. Analysis of the headspace above chopped garlic using GC-AED shows MeS_nMe, MeS_nAll, and $AllS_nAll$ ($n = 1-3$, All = allyl) in the S channel. The Se channel shows seven peaks: dimethyl selenide (MeSeMe), methanesulfenoselenoic acid methyl ester (MeSeSMe), dimethyl diselenide (MeSeSeMe), bis(methylthio)selenide (($MeS)_2Se$), allyl methyl selenide (MeSeAll), 2-propenesulfenoselenoic acid methyl ester (MeSeSAll), and (allylthio)(methylthio)selenide (MeSSeSAll).[22,23] Structures were established by GC-MS using synthetic standards. The headspace above chopped onion contained methyl propyl selenide, MeSePr.

Lyophilized normal garlic (0.02 ppm Se) or moderately Se-enriched garlic (68 ppm Se; grown in Se-enriched soil) was derivatized with ethyl chloroformate to volatilize the selenoamino acids, likely precursors of the headspace Se compounds. Analysis by GC-AED showed selenocysteine, identified by comparison with the mass spectral fragmentation and the retention time of an authentic standard. In more heavily Se-enriched garlic (1,355 ppm Se), Se-methyl selenocysteine was the major selenoamino acid found along with minor amounts of selenocysteine and traces of selenomethionine; the S channel showed 2 : 1 allyl-cysteine and allyl-cysteine *S*-oxide along with minor amounts of methionine.[24] There were only minor changes in the ratios of the sulfur amino acids as the level of Se was varied from 0.02 to 1,355 ppm. Similar analysis of Se-enriched onion (96 ppm Se) revealed the presence of equal amounts of *Se*-methyl selenocystein and selenocysteine in the Se channel.

HPIC as well as C18 LC with ICP-MS detection has been used to analyze the Se species in garlic, onion, and yeast without the application of heat.[25-27] These analyses show that Se-methyl selenocysteine is the major component along with lesser amounts of Se-methione, selenocystein, and selenate and se-

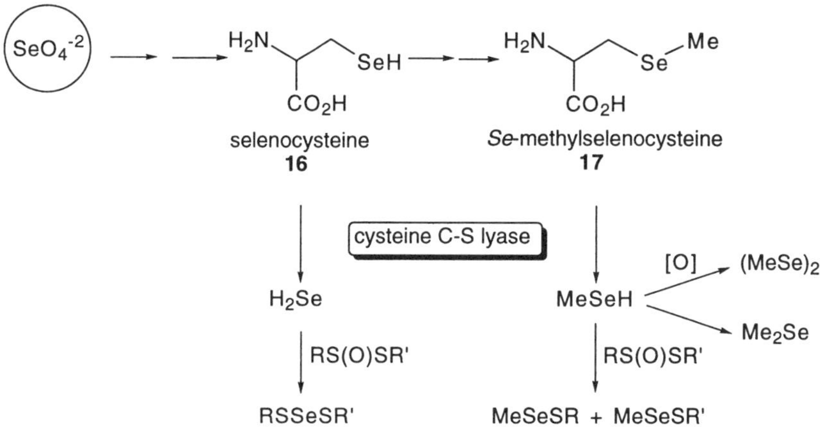

Scheme 7

lenite salts in high-Se garlic. However, a considerable number of interesting unknown Se-containing components remain to be identified. A proposed mechanism for formation of the various selenium compounds thus far identified from garlic is shown in Scheme 7.

TRANSFORMATION OF *ALLIUM* FLAVORANTS FOLLOWING COOKING AND INGESTION

Research on the transformation of *Allium* flavorants following cooking and ingestion has been mainly limited to garlic. Alliinases are completely inactivated when unpeeled whole garlic cloves are boiled for 15 to 20 minutes.[1] However, before completion of this process, up to 1% of the precursor alliin is converted into allicin, perhaps by mechanical abrasion of the cloves during boiling.[1] Boiling rapidly converts any allicin formed into diallyl trisulfide and related polysulfides, which are detected in breath following ingestion of boiled garlic. At the same time, peptide precursors of alliin (e.g, γ-glutamyl-cysteines) are hydrolyzed to *S*-allylcysteine and *S*-1-propenylcysteine, which are further degraded on continued boiling.[1] If garlic is crushed prior to boiling, most alliin is converted to allicin and other thiosulfinates; boiling for approximately 20 minutes completely transforms these into polysulfides. Boiling cut garlic in water in an open container leads to loss of 97% of volatile sulfides by evaporation. However, even at the high temperatures (approximately 180°C) involved in stir-frying chopped garlic cloves in hot soybean oil for 1 minute, 16% of the sulfides are retained in the oil.[1] Denaturization of alliinase occurs particularly rapidly on microwaving: complete inactivation of alliinase in individual 5- to 6-gram garlic cloves occurs within 15 to 30 seconds (65 watts microwave power).[1]

What happens to garlic flavorants upon ingestion? Our analysis of human garlic breath by GC-AED (the subject consumed, with brief chewing, 3 g of fresh garlic with small pieces of white bread, followed by 50 mL of cold water) showed in the Se channel dimethyl selenide (MeSeMe) as the major Se component along with smaller amounts of $MeSeC_3H_5$, MeSeSMe, and $MeSeSC_3H_5$; the S channel showed AllSH, MeSAll, and AllSSAll with lesser amounts of MeSSMe, $MeSSC_3H_5$, an isomer of AllSSAll (presumably MeCH = CHSSAll), $C_3H_5SC_3H_5$, and $C_3H_5SSSC_3H_5$.[28,29] In this same study, we also examined the composition of the Se and S compounds in garlic breath as a function of time. After 4 hours, the levels of MeSeMe, AllSSAll, AllSAll, and MeSSMe were reduced by 75% from the initial levels of 0.45 ng/L (MeSeMe), 45 ng/L (AllSSAll), 6.5 ng/mL (AllSAll), and 1.8 ng/L (MeSSMe). The AllSH could only be found in breath immediately after ingestion of garlic. In view of the reported very low threshhold detection level for low molecular weight organoselenium compounds,[30] it is likely that compounds such as MeSeMe contribute to the overall odor associated with garlic breath. It has been previously reported that MeSeMe, which has a garlic-like odor, is found in the breath air of animals fed inorganic Se compounds[31] and humans who have accidentally over ingested Se compounds.[32] Studies involving consumption of larger quantities of garlic (38 g) indicate persistence of levels of S compounds as high as 900 ppb in the subject's breath for more than 32 hours.[33] The presence of elevated levels of acetone in the subject's breath was attributed to enhanced metabolism of blood lipids.[33]

COMMERCIALLY AVAILABLE GARLIC PREPARATIONS

In addition to dried or pickled forms of garlic sold in food stores, garlic health supplements are available in the following forms: (1) powdered tablets or capsules containing dehydrated garlic powder, consisting of alliin and related flavor precursors, the enzyme alliinase, and cysteine-containing peptides. On exposure to water, this product yields up to 5 mg/g of allicin (the "allicin-potential"). (2) Steam-distilled oil of garlic, consisting primarily of allyl polysulfides. (3) Garlic vegetable oil macerate (garlic crushed in oil), containing ajoene, dithiins, and allyl polysulfides. (4) Garlic extract aged in dilute alcohol, consisting mainly of S-allyl-cysteine and cysteine-containing peptides; no allicin is formed on exposure to water. There is significant variation in the quantities of the various garlic components in different commercial products. Currently, there are no requirements for standardization or for labeling of components or their quantities on product labels.

SUMMARY OF *ALLIUM* BIOLOGICAL ACTIVITY

Allicin and related thiosulfinates are known to possess antibacterial, antifungal, and antitumor activity.[1,2] Allicin also displays lipid biosynthesis inhibition and antithrombotic activity.[1,2] Both ajoene and cepaenes have antithrombotic activity; ajoene also displays antifungal activity.[1,2] *Allium* thiosulfinates (as well as ajoene) react readily with thiols such as cysteine and undergo hydrolysis. These facts severely limit the *in vivo* lifetimes of ajoene and *Allium* thiosulfinates. Thus, metabolites of these compounds are more likely to be the active agents *in vivo*. Allicin precursor γ-glutamyl *S*-allyl cysteine inhibits the blood pressure-regulating angiotensin-converting enzymes, while the MeS(O)SMe (**3c**) precursor *S*-methyl cysteine *S*-oxide (**1c**) inhibits the formation of benzo[a]pyrene-induced micronucleated polychromatic erythrocytes (an indicator for genotoxicity).[2] In other laboratory studies, allylic sulfides and selenides are reported to inhibit growth of tumor cells.[2] Recent epidemiological studies suggest correlations between consumption of fresh garlic and decreased risk of gastrointestinal cancer[34] and occurrence of colorectal polyps.[35]

REFERENCES

1. Koch, H. P., and Lawson, L. D., (eds.). 1996. *Garlic. The Science and Therapeutic Application of* Allium sativum *L. and Related Species.* 2nd Ed. Baltimore, MD: Williams & Wilkins, 329.

2. Block, E. 1996. The chemistry and health benefits of organosulfur and organoselenium compounds in garlic (*Allium sativum*): recent findings, In: *Hypernutritious Foods.* (Finley, J. W., Armstrong, D. J., Nagy, S., Robinson, S. F., eds.). Auburndale, FL: Agscience, pp. 261–292.

3. Block, E. 1992. The organosulfur chemistry of the genus *Allium*-implications for organic sulfur chemistry, *Angew. Chem. Int. Ed. Engl.* 31:1135–1178.

4. Cavallito, C. J., Bailey, J. H., Buck, J. S. 1945. The antibacterial principle of *Allium sativum.* III. Its precursor and essential oil of garlic, *J. Am. Chem. Soc.,* 67:1032–1033.

5. Block, E., Ahmad, S., Catalfamo, J., Jain, M. K., Apitz-Castro, R. 1986. Antithrombotic organosulfur compounds from garlic: structural, mechanistic and synthetic studies, *J. Am. Chem. Soc.* 108:7045–7055.

6. Block, E., 1993. Flavor Artifacts, *J. Agric. Food Chem.,* 41:692.

7. Block, E., Calvey, E. M. 1994. Facts and artifacts in *Allium* chemistry. In: *Sulfur Compounds in Foods,* ACS Symposium Series 564 (Mussinan, C. J., and Keelan, M. E. eds.), Washington, DC: American Chemical Society, pp. 63–79.

8. Calvey, E. M., Betz, J. M., Matusik, J. E., White, K. D., Block, E., Littlejohn, M. H., Naganathan, S., Putman, D. 1994. Off-line supercritical fluid extraction of thiosulfinates from garlic and onion, *J. Agric. Food Chem.* 42:1335–1341.

9. Calvey, E. M., Block, E. 1997. Supercritical fluid extraction of *Allium* species, in *Spices: Flavor Chemistry and Antioxidant Properties,* ACS Symposium Series 660, Risch, S. J., and Ho, C.-T. eds., American Chemical Society: Washington DC; pp. 113–124.

10. Block, E., Putman, D., Zhao, S.-H. 1992. *Allium* chemistry: GC-MS Analysis of thiosulfinates and related compounds from onion, leek, scallion, shallot, chive and Chinese chive, *J. Agric. Food Chem.* 40:2431–2438.

11. Block, E., Naganathan, S., Putman, D., Zhao, S.-H. 1992. *Allium* chemistry: HPLC quantitation of thiosulfinates from onion, garlic, wild garlic, leek, scallions, shallots, elephant (great-headed) garlic, chives and Chinese chives. Uniquely high allyl to methyl ratios in some garlic samples," *J. Agric. Food Chem.* 40:2418–2430.

12. Calvey, E. M., Matusik, J. E., White, K. D., DeOrazio, R., Sha, D., Block, E. 1997. *Allium* chemistry: supercritical fluid extraction and LC-APCI-MS of thiosulfinates and related compounds from homogenates of garlic, onion and ramp. Identification in garlic and ramp and synthesis of 1-propanesulfinothioic acid *S*-allyl ester, *J. Agric. Food Chem.* pp. 4406–4413.

13. Block, E., Bayer, T., Naganathan, S., Zhao, S.-H. 1996. *Allium* chemistry: synthesis and sigmatropic rearrangements of alk(en)yl 1-propenyl disulfide *S*-oxides from cut onion and garlic, *J. Am. Chem. Soc.* 118:2799–2810.

14. Block, E., Thiruvazhi, M., Toscano, P. J., Bayer, T., Grisoni, S., Zhao, S.-H. 1996. *Allium* chemistry: structure, synthesis, natural occurrence in onion (*Allium cepa*), and reactions of 2,3-dimethyl-5,6-dithiabicyclo[2.1.1]hexane *S*-oxides, *J. Am. Chem. Soc.* 118:2790–2798.

15. Shakespeare, W. 1623. *Antony and Cleopatra,* I, ii, 173.

16. Block, E., Gillies, J. Z., Gillies, C. W., Bazzi, A. A., Putman, D. Revelle, L. K., Wall A., Wang, D., Zhang, X. 1996. *Allium* chemistry: microwave spectroscopic identification, mechanism of formation, synthesis, and reactions of (*E,Z*)-propanethial *S*-oxide, the lachrymatory factor of the onion (*Allium cepa*), *J. Am. Chem. Soc.* 118:7492–7501.

17. Schmidt, N. E., Santiago, L. M, Eason, H. D., Dafford, K. A., Grooms, C. A., Link, T. E., Manning, D. T., Cooper, S. D., Keith, R. C., Chance, W. O., III, Walla, M. D., Cotham, W. E. 1996. A rapid extraction method of quantitating the lachrymatory factor of onion using gas chromatography, *J. Agric. Food Chem.* 44:2690–2693.

18. Penn, R. E., Block, E., Revelle, L. K. 1978. Methanesulfenic acid, *J. Am. Chem. Soc.* 100:3622–3623.

19. Spåre, C. G., Virtanen, A. I. 1964. On the occurrence of free selenium-containing amino acids in onion (*Allium cepa*), *Acta Chem. Scand.* 18:280–282.

20. Ip, C., Lisk, D. J., Stoewsand, G. S. 1992. Mammary cancer prevention by regular garlic and selenium-enriched garlic, *Nutr. Cancer* 17:279–286.

21. Clark, L. C. et al. 1996. Effects of selenium supplementation for cancer prevention in patients with carcinoma of the skin, *J. Am. Med. Assoc.* 276:1957–1963.

22. Cai, X.-J., Uden, P. C., Sullivan, J. J., Quimby, B. D., Block, E. 1994. Headspace/gas chromatography with atomic emission and mass selective detection for the determination of organoselenium compounds in elephant garlic, *Anal. Proc. Including Anal. Commun.* 31:325–327.

23. Cai, X.-J, Uden, P. C., Block, E., Zhang, X., Quimby, B. D., Sullivan, J. J. 1994. *Allium* chemistry: identification of natural abundance organoselenium volatiles from

garlic, elephant garlic, onion, and Chinese chive using headspace gas chromatography with atomic emission detection, *J. Agric. Food Chem.* 42: 2081–2084.

24. Cai, X.-J., Block, E., Uden, P. C., Zhang, X., Quimby, B. D., Sullivan, J. J. 1995. *Allium* chemistry: identification of selenoamino acids in ordinary and selenium-enriched garlic, onion, and broccoli using gas chromatography with atomic emission detection, *J. Agric. Food Chem.* 43:1754–1757.

25. Ge, H., Tyson, J. F., Uden, P. C., Xai, X.-J., Denoyer, E. R., Block, E. 1996. Identification of selenium species in selenium-enriched garlic, onion, and broccoli using high-performance ion chromatography with inductively coupled plasma mass spectrometry detection, *Anal. Commun.* 33:279–281.

26. Bird, S. M., Ge, H., Uden, P. C., Tyson, J. F., Block, E., Denoyer, E. R. 1998. High performance liquid chromatography of selenoamino acids and organo-selenium compounds: speciation by inductively coupled plasma mass spectrometry (HPLC-ICP-MS), *J. Chromatog.*, 12:785–788.

27. Bird, S. M., Uden, P. C., Tyson, J. F., Block, E., Denoyer, E. R. Speciation of selenoamino acids and organo-selenium compounds in selenium-enriched yeast using high performance liquid chromatography inductively coupled plasma mass spectrometry, *Anal. Commun.*, in press.

28. Cai, X.-J., Block, E., Uden, P. C., Quimby, B. D., Sullivan, J. J. 1995. *Allium* chemistry: identification of natural abundance organoselenium compounds in human breath after ingestion of garlic using gas chromatography with atomic emission detection, *J. Agric. Food Chem.* 43:1751–1753.

29. Block, E., Cai, X.-J., Uden, P. C., Zhang, X., Quimby, B. D., Sullivan, J. J. 1996. *Allium* chemistry: natural abundance organoselenium compounds from garlic, onion, and related plants and in human garlic breath, *Pure Appl. Chem.* 68:937–944.

30. Ruth, J. H. 1986. Odor thresholds and irritation levels of several chemical substances: A review. *Am. Ind. Hyg. Assoc. J.* 47:142–151.

31. Oyamada, N., Kikuchi, M., Ishizaki, M. 1987. Determination of dimethyl selenide in breath air of mice by gas chromatography. *Anal. Sci.* 3:373–376.

32. Buchan, R. F. 1974. Garlic Breath Odor. *J. Am. Med. Assoc.* 227:559–560.

33. Taucher, J., Hansel, A., Jordan, A., Lindinger, W. 1996. Analysis of compounds in human breath after ingestion of garlic using proton-transfer-reaction mass spectrometry. *J. Agric. Food Chem.* 44:3778–3782.

34. Steimetz, K. A., Kushi, L. H., Bostick, R. M., Folsom, A. R., Potter, J. D. 1994. Vegetables, fruit, and colon cancer in the Iowa Women's Health Study. *Am. J. Epidemiol.* 139:1–15.

35. Witte, J. S., Longnecker, M. P., Bird, C. L., Lee, E. R., Franid, H. D., Haile, R. W. 1996. Relation of vegetable, fruit and grain consumption to colorectal adenomatous polyps. *Am. J. Epidemiol.* 144:1015–1025.

CHAPTER 9

Emerging Applications of Fungal Chemistry

MARTIN F. STONER

INTRODUCTION

$\mathbf{F}$UNGI in the phyla Ascomycota (e.g., yeasts, morels, and truffles), Basidiomycota (e.g., mushrooms and polypores), and Zygomycota (e.g., *Absidia, Mucor,* and *Rhizopus*) and asexual forms of the former two groups (e.g., *Aspergillus* and *Penicillium,* basidiomycetous yeasts) comprise a vast and only slightly tapped resource of biochemicals applicable to the savory character, nutritional value, and healthfulness, utility, and processing of foods, dietary supplements, and related products (Alexopoulos et al., 1996; Beuchat, 1987; Onions et al., 1986).

In addition to accumulating scientific knowledge, there is a long history of uses of fungi for nutrition and health in the East and West (Alexopoulos et al., 1996; Benjamin, 1995; Chang, 1972; Stamets, 1993; Ying et al., 1987) to guide scientists and technologists in the quest for new physiologically active fungal chemicals and applications. Current knowledge on the chemical interactions of fungi and plants constitutes a broad reference base for the development of new uses and products involving fungus-ripened foods, by-products, and incorporation of fungus tissues or novel secondary chemicals. New applications of fungal biochemistry and previously impractical ones are being facilitated and expanded via genetic engineering and other biotechnologies (Griffin, 1994; Murasugi et al., 1991; Pollock, 1992). Traditional and emerging uses of fungi and fungal chemistry for nutritional, disease-preventative, therapeutic, and other health-related purposes are being validated or expanded through vigorous and

143

varied research involving fungi such as the ascomycete *Cordyceps* (Alexopoulos et al., 1996; Kiho and Ukai, 1995) and Basidiomycetes such as the shiitake mushroom, *Lentinus edodes* (Alexopoulos et al., 1996; Block, 1994; Jong and Birmingham, 1993) and the Ling Zhi fungus, *Ganoderma lucidum* (Alexopoulos et al., 1996; Shiao et al., 1994) as well as other wood-decaying species and microfungi. Traditionally assumed and scientifically proven beneficial effects of fungal chemistry in human health are shown in Table 1. Special attention is called to the physiological actions of fungal polysaccharides, proteins, triterpenes, and/or antibiotics in the prevention or therapy of cancer, viral infections, and other diseases; in immunomodulation; in inhibition of cholesterol synthesis; and in general wellness. Many fungal biochemicals have been isolated, identified, and employed in research; the genes for some compounds have been biotechnically characterized, cloned, or transplanted. Some health benefits of fungal chemicals may be obtained now via dietary or medicinal products (Jong and Birmingham, 1993), and the literature points to suitable fungi for further investigation. Many of these fungi can be cultivated now under controlled conditions for maximal productivity (Stamets, 1993).

TABLE 1. Beneficial Effects of Fungal Chemistry in Human Health. An Increasing Number of Benefits Formerly Indicated Only by Traditional Medicine Join the Ranks of Actions Confirmed by Medical Research.

Confirmed by Medical and Nutritional Research	Strongly Suggested by Traditional Medicine[a]
Antibiotic[b]	Alleviate swelling
Antitoxicant[c]	Antiinflammatory
Antitumor[d]	General tonic
Healing Aid[e]	Improve digestion
Immunomodulation[d]	Prevent bleeding[f]
Physiologic regulation[d]	Prevent bloating
Psychoactive[g]	
Vitamins, nutrients[h]	

[a]Ying et al., 1987.
[b]Alexopoulos et al., 1996; Benjamin, 1995; Garroway and Evans, 1984; Jong and Birmingham, 1993; Quack et al., 1978.
[c]Xu et al., 1995.
[d]Jong and Birmingham, 1993; Shiao, Lee, et al., 1994.
[e]Vanderhem et al., 1994.
[f]Gilbertson, 1980.
[g]Lincoff and Michell, 1977.
[h]Benchat, 1987; Robinson, 1987.

Considering the well-established and vast base of available, edible fungi, possibilities abound for new applications of fungal chemistry to improvement of food products via flavor and textural enhancements without significant caloric increases; increased protein, essential fatty acid, and vitamin content; incorporation of unique health-promoting fungal chemicals in foods; and fungal enzyme modification of biochemicals from plants or animals. Of particular interest are fungal exoenzymes, lectins and other proteins, polysaccharides, terpenoids, sugars, and flavor or fragrance compounds.

Interactions of fungi with phytochemicals could produce altered or augmented phenolic and terpenoid compounds or other biochemicals of potential value in nutrition and health. Given the ability of fungi to reduce plant biomass and produce unique and useful biochemicals, integration of fungal processes in food and nutritional supplement manufacturing could improve utilization of organic inputs and reduce waste while increasing the diversity and value of commercial products.

ECOLOGICAL, EVOLUTIONARY, AND BIOCHEMICAL ATTRIBUTES OF FUNGI RELATIVE TO HUMAN NUTRITION AND HEALTH

To best appreciate the great potential for applications of fungal chemistry, and to more easily envision possible resources and avenues for research and development, it is necessary to examine the natural systems that form the well-developed basis for diverse chemical interactions among humans, fungi, and plants.

The involvement and value of fungi in human nutrition and health is not accidental; it appears to be rooted in evolutionary, ecological, and biochemical organization, with humans being more than casual partners. Essential, evolved microbial interfaces affecting nutrition, health, and overall functioning of individuals and their ecosystems (Stoner, 1994) exist among humans and plants and their environment (Figure 1). Fungi are major partners in these interfaced systems. Within such ecosystems, diversity in biochemistry (food substrates and other metabolites) supports biodiversity; in turn, biodiversity supports health, functional efficiency and integrity, and survival (Agrios, 1997; Alexopoulos et al., 1996; Stoner, 1994). Therefore, practical applications of fungal chemistry to human nutrition and health can be based on and derived from natural systems and materials and are a very promising area for research and development. This organization and its interactive biochemical systems constitute a foundation for very practical developments in the multifaceted utilization of fungi in human nutrition and health and in sustainable/regenerative systems (Figure 2).

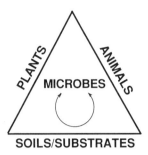

Figure 1 Microbes serve as essential interfaces in nutrient cycling and other life-mediating and ecosystem-integrating processes.

HUMAN PERSPECTIVES

Over the millennia, our ancestors have found nourishment, beauty, pleasure, spiritual communion, cultural traditions, and medicinal benefits in fungi (Alexopoulos et al., 1996; Benjamin, 1995; Christensen, 1975; Stamets, 1996; Ying et al., 1987). The use of fungi in nutrition and medicine evidently was developing thousands of years ago in the East and West (Alexopoulos et al., 1996; Jong and Birmingham, 1992, 1993; Ying et al., 1987). *Shen Nong's Herbal,* in first century B.C. China, listed medicinal effects of several fungi (Ying et al., 1987). Human ties with fungi could have been cemented even longer ago by co-evolutionary developments and attendant recognition systems established among the biochemistries of our primate ancestors, plants and fungi.

As mentioned above, ancient societies revered fungi for the special qualities they brought to human life. We can presume that in those societies the use of fungi for food (mycophagy), psychoactive or spiritual uses, medicine,

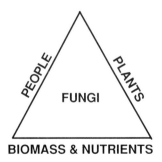

Figure 2 The fungal interface, with its potential interchanges among plants, biomass and nutrients, and people, facilitates natural and industrial processes in support of human nutrition and health in a sustainable environment.

or other purposes was considered part of daily life or ritual and was available freely or with guidance to the populace (Benjamin, 1995; Stamets, 1996). In countries such as China and Japan, where many aspects of ancient culture have been conserved, the holistic uses of fungi in food and medicine are still looked upon as a normal part of life (Benjamin, 1995; Jong and Birmingham, 1993; Ying et al., 1987); such diversity in the diet could contribute to wellness. However, in societies where geographic displacement of people, immigration, and cultural mixing and homogenization have occurred, traditional uses of and consciousness about fungi often have faded or disappeared. Indeed, in areas such as England and the United States, the passage of time and consequent lack of familiarity with traditional uses of fungi as food and medicine has given way to profound ignorance or even fear of fungi ("mycophobia") (Benjamin, 1995) and resultant losses of potential benefits in holistic nutrition and health. This is one facet of a societal condition that I have called "cultural deficiency syndrome." Slowly but surely, aided by expanding global communications and cultural interaction, more and more people are rediscovering holistic nutrition and health and, naturally, the fungi.

People, of course, do have good reason to be cautious about consuming certain fungi. Poisonous chemicals ("mycotoxins"), including some potentially lethal and/or carcinogenic substances such as the aflatoxins, amatoxins, ergotine alkaloids, monomethylhydrazine, and trichothecenes, are found in some Ascomycetes and Basidiomycetes (Beuchat, 1987; Christensen, 1975; Lincoff and Mitchell, 1977; Stamets, 1996). Still, these fungi and their mycotoxins are well documented and can be avoided in daily life or research and development by proper education, testing, and adherence to food content and handling standards and regulations.

BIOCHEMICAL AFFINITIES AMONG FUNGI, PLANTS, AND HUMANS

Particularly apropos to our interests in phytochemistry, food, and health is the unique position of fungi in the biochemical and ecological interfaces between people and plants. Plants are very important in the nutrition of people *and* fungi; fungi commonly and often conspicuously inhabit areas where people gather and/or process plants and plant products.

Fungi are heterotrophs, which, by existing as parasites, commensals, or saprotrophs, absorb nutrition from living or dead organisms and their decomposition products (Agrios, 1997; Alexopoulos et al., 1996; Cooke and Rayner, 1984; Cooke and Whipps, 1993; Garraway and Evans, 1984). The fungi can utilize a vast variety of organic substances as carbon and/or energy sources (Cooke and Whipps, 1993; Garraway and Evans, 1984). Ascomycetes and Basidiomycetes as groups possess pronounced enzymatic diversity. Although a number of fungi utilize nutrients from animals or even other microbes, most prefer, and are evolutionarily adapted to, the partnership and biochemical resources of plants. Nutrition is obtained by fungi from plants via saprotrophism, parasitism, or staged combinations of both avenues (Agrios, 1997; Cooke, 1977; Cooke and Rayner, 1984). Since biochemical-nutritional dependencies and fungal lifestyles have been laced together with a myriad of evolutionary adaptations in morphological, physiological, and ecological characteristics, fungi tend to occur together with plants in predictable, functionally integrated communities in ecosystems (Alexopoulos et al., 1996; Cooke, 1977; Cooke and Rayner, 1984; Stoner and Baker, 1981). Such associations over the ages have led to numerous biochemical similarities, interfaces, and exchanges between fungi and plants. Such biochemical interactions and systems are most developed in plant host-parasite relationships (Agrios, 1997).

Human physiology is inherently receptive to fungal and plant nutrients and is reactive to certain metabolic products of both (possibly via lectin-based recognition systems and other predispositions) (Benjamin, 1995; Lincoff and Michell, 1977).

FUNGAL BIOCHEMICAL PATHWAYS AND PRODUCTS OF SPECIAL INTEREST

Because of their growing reputation for both utility in biotechnical processes and fascinating secondary chemistry, certain fungal biochemical pathways and associated enzymes, intermediates, or products (Figure 3) continue to attract scientific and industrial interest. Many of these pathways or metabolites have close relationships with biochemical systems and intermediates in plants (Garraway and Evans, 1984; Geissman and Crout, 1969; Griffin, 1994; Robinson, 1980; Weete, 1974), thereby providing grounds for interaction of plant and fungal chemistries in nature or industrial processes. Various fungi possess a broad array of enzymes or groups of enzymes useful in industry, including amylases, glucosidases, cellulases, chitinases, ligninases, lipases, pectinases, peroxidases, and phenoloxidases (Bigelis and Lasure, 1987; Bonnen, 1994; Cooke and Whipps, 1993; Garraway and Evans, 1984; Griffin, 1994; Weete, 1974; Zhang, 1997). Noteworthy among the pathways are those concerned with diverse phenolic metabolism and alkaloid synthesis (Cooke and Whipps, 1993; Garraway and Evans, 1984; Griffin, 1994), glycolysis and the pentose phosphate pathways (Cooke and Whipps, 1993; Garraway and Evans, 1984), terpenoid pathways (Weete, 1974), the shikimate pathway (Garraway and Evans, 1984), polyketide pathways (Garraway and Evans, 1984), and the enzyme complexes associated with degradation or synthesis of cellulose, lignin, and other polyphenols (Agrios, 1997; Bonnen, 1994; Cooke and Whipps, 1993; Griffin, 1994; Thurston, 1994; Zhang, 1997). Specific chemical groups of much interest include the fungal polysaccharides, isoprenoids, triterprenes (Hirotana, 1993; Tai 1993; Weete, 1974), diverse

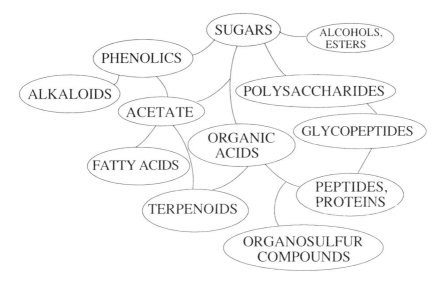

Figure 3 Principle pathways and products contributing to special nutritive and bioactive properties of fungi.

phenolics and derivative aromatic amino acids (Garraway and Evans, 1984; Griffin, 1994), lectins (Conrad, 1994; Kawagishi, 1997; Kiho and Ukai, 1995), and bioactive nucleosides (Kiho and Ukai, 1995; Kuo et al., 1994; Shiao et al., 1994). Fungal proteases have been useful in some applications, but may not be economical choices in others (Beuchat, 1987; Bigelis and Lasure, 1987). Attention has been called to the possibility that auxiliary ("dirigent") proteins affecting biomolecular phenoxy radical coupling (as shown in plants) could occur in fungal fruiting bodies (Davin et al., 1997). This could be of special interest in understanding the health-promoting qualities and product potential of fungal chemistry.

Lectins are proteins that have special affinities with polysaccharides and that appear in some cases to have roles in biological recognition and interaction among organisms as well as modulating influences on processes in normal or pathological physiology and disease defense (Peumanss, 1995; Robinson, 1980). Lectins occur in various fungi, including *Pleurotus* (Conrad, 1994) and the genera *Cordyceps, Ganoderma,* and *Lentinus,* discussed herein under Health-Promoting Attributes of Fungal Chemistry (Jong and Birmingham, 1993; Kawagishi, 1997; Kiho & Ukai, 1995). They may play important roles in some immunomodulatory or other actions of fungi.

Taxol, a chemical used in the treatment of human ovarian cancer, and originally derived from the Pacific Yew tree (*Taxus*), is produced by the fungus *Taxomyces andreanae* (Stierle, 1993). While it is questionable whether this particular fungus could be used for commercial production, its taxol-production genes could be transplanted to more industrially manageable fungi or bacteria in the future.

Some fungal chemicals such as muscarine and psilocybin found in some mushrooms have distinct psychoactive properties such as soporific, depressent, or hallucinatory actions (Lincoff and Michell, 1977; Stamets, 1996). It is conceivable that such compounds or derivatives might someday have significant application to human mental health.

Antibiotics produced by fungi include nucleosides such as *Cordycepin* (Kiho and Ukai, 1995; Kuo et al., 1994), cinnamic acid derivatives, nonaromatic amino acid derivatives such as penicillins and cephalosporins, aromatic amino acid derivatives, polyacetylenes from *Aspergilli,* cyclic peptides such as the *Amanita* toxins, hybrids of aromatic amino acids and terpenoids such as lysergic acid amides, and polyketides such as patulin and the derivative grifolin (Garraway and Evans, 1984; Lincoff and Michell, 1977). Certain antibiotic-flavorant, dipeptide organosulfur compounds found in "garlic" (*Marasmius*) and "shiitake"(*Lentinus*) mushrooms (Block, 1994; Block et al., 1994; Gmelin et al., 1976, 1980; Pyysalo, 1975; Rapior, 1997) are chemically similar to compounds in *Allium* and therefore warrant more study regarding potential health-promoting values as well as savory contributions to food products.

FUNGAL CHEMISTRY IN NUTRITION AND HEALTH

Today there is an elevated consciousness of food and its relationship to health. Furthermore, it is known that natural chemicals ingested as part of food can have health-promoting or diminishing effects. Such chemicals could be interacting with body systems to prevent pathogenic processes while others could be serving to inhibit or perhaps even cure already existing pathologies, such as cancers. Based on emerging research results, corroborated in many cases by centuries of traditional medicine (Benjamin, 1995; Ying et al., 1987), this appears to be the case also with bioactive fungal chemicals.

Although some of the fungal chemicals discussed here have been used or studied primarily as medicines, attention to their chemistry and properties should provide insights to potential uses as beneficial food constituents or supplements that could exert benefits when ingested under a healthful dietary and lifestyle regime. Some of the fungi mentioned herein as sources of compounds for use in medicine, such as the shiitake mushroom. *L. edodes* (Jong and Birmingham, 1993; Ying et al., 1987), have already been used for centuries by millions of people as part of their normal diet. Physiologically active compounds in such fungi are ingested regularly in small amounts, and they may be making incremental contributions, together with many other factors, to wellness.

CONTRIBUTIONS TO KNOWLEDGE FROM TRADITIONAL AND SCIENTIFIC MEDICINE

Prior to considering specific examples of attributes and applications of fungal chemistry in human health, it is important to develop certain perspectives regarding the relative natures of traditional and scientific medicine as they relate to this subject, the contributions of each to this topic, and the interrelationships of both in ongoing research. Awareness and understanding in these areas facilitates more imaginative and diverse approaches to research, development of uses and products, and marketing in the food industries as well as in pharmaceuticals.

In developing such perspectives, it is counterproductive and confusing to use the terms "Eastern medicine" and "Western medicine," although they have been employed in contrasting, respectively, nonscientific, traditional, and scientific approaches to the development of knowledge and applications in medicine. All cultures, both Eastern and Western, have elements of traditional medicine in their backgrounds; today, both in the East and West, elements of the traditional and scientific approaches are considered valuable and in certain ways are merging in the quest for knowledge and effective applications in human health.

Traditional medicine, as practiced in China where extensive written records have been kept for centuries, "emphasizes whole-body homeostasis instead of (simply) disease-oriented therapy. Food and herbal drugs are used together for the prevention of illness" (Shiao et al., 1994). In addition to food and herbal drugs, mineral and animal materials as well as mental (e.g., meditation) and physical exercise regimens may be prescribed for holistic health. The contrasting scientific medicine emphasizes research on individual aspects and direct treatments of specific pathogens or pathologies, frequently involving chemically defined pharmaceuticals, sometimes without emphasis on whole-body conditions (Benjamin, 1995). In the treatment of diseases wherein both approaches are effective, traditional (whole-body) remedies may be slower in action but are generally noted for having fewer or no side effects, whereas specifically targeted drug treatments in scientific medicine may be quick-acting but may involve significant side effects (Benjamin, 1995; Xu et al., 1995; Ying et al., 1987). Such side effects are believed to be due to the artificial nature of many modern drugs, whereas the drugs in traditional medicine are derived from complex natural materials, some of which may act as potentiators against possible "imbalances" or side effects by other components. Today medical treatments in both the East and West tend to combine beneficial elements of both traditional and scientific medicine. Interestingly, extracts from *Cordyceps sinensis,* a traditional medicine fungus, have been shown to ameliorate cyclosporine drug-induced nephrotoxicity in kidney transplant recipients (Xu et al., 1995).

Both traditional and scientific medicine have made significant contributions in identifying valuable medicinal fungi and/or in elucidating the chemical and pharmacological bases for beneficial effects. Positive correlations between reported benefits for certain fungi or extracts therefrom in traditional medicine and those validated for the same organisms by scientific research are strongest in cases in which specific fungal compounds acting directly against certain pathogens or pathological mechanisms are involved (Benjamin, 1995). Examples of such validations can be found in many recent scientific articles about *L. edodes* (shiitake) (Jong and Birmingham, 1993), *Cordyceps* (Chan Hua) (Kiho and Ukai, 1995; Kuo et al., 1994), and *G. lucidum* (Ling Zhi or Reishi fungus) (Shiao et al., 1994). Of course, the medical values of certain fungal drugs investigated in the past, such as penicillin, ergotine alkaloids, and cephalosporins, already have been well proven (Alexopoulos et al., 1996; Garraway and Evans, 1984).

Progress has been slower and more difficult in scientifically elucidating and validating health benefits reported in traditional medicine, wherein apparently combined effects of various chemical agents from singular or mixed sources are involved. Many prescriptions in traditional Chinese medicine, for example, involve more than one material (plant, animal, or mineral) and may involve items that are generally recognized as either food or medicine. Indeed, in many cases, "there is no strict borderline between drug treatment and dietary manipulation" (Shiao et al., 1994). This holistic application of materials has been called "polypharmacy" (Benjamin, 1995) in contrast to the prescription of singular drugs. In the polypharmacy approach, chemical and pharmacological bases of action often are not understood. In such cases, scientific research to elucidate active multiple ingredients and to validate traditionally recognized effects is slow, complex, and difficult. Nevertheless, persistent research by numerous scientists appears to be making significant progress in this area, as exemplified by studies on the shiitake mushroom, wherein different biochemicals and allied physiological actions have been scientifically validated and related in part to traditional uses (Jong and Birmingham, 1993).

Progress in this area of research has generated intense interest in the value of fungal chemistry in human health, as evidenced by burgeoning scientific publications around the world as well as by increasing pharmaceutical and food supplement applications of such fungi (Jong and Birmingham, 1992, 1993). The situation is reminiscent of the explosion of interest in medicinal plant chemistry that occurred early in this century, which is rekindled today by increasing consciousness about biodiversity and the values of phytochemisty in holistic medicine.

The combination of dietary and medicinal fungi in traditional medicine, as well as the established edibility of many fungi, point strongly to potential uses of fungal components in both food supplements as well as pharmaceuticals in future product development. A model for this area of development is inherent in the utilization of shiitake (*L. edodes*) (Jong and Birmingham, 1993).

Problems that have hindered research on fungal chemicals in human health have included the lack of awareness of fungi in some parts of the world; the relatively recent nature of much scientific research—particularly on a cooperative, global scale; language barriers and consequent lack of awareness and broad reinforcement of research; the challenges inherent in studying complex systems of chemistry and actions, including those involving slow-acting or weakly defined beneficial effects; and lack of dosage trials and/or extensive testing with human subjects. Increasing global awareness and participation in research are working to address and alleviate these problems.

Major challenges that remain include developing more effective research linkages between traditional medicine and scientific medicine worldwide for identifying and evaluating fungal chemicals; engaging and guiding the pharmaceutical and food industries in creative strategies for the development of products using fungal chemistry; and developing better policies and systems for regulating products and their production in ways that maintain assurances of high quality, effectiveness, and safety while at the same time better serving the broadening concepts and demands of food science and medicine.

Denis Benjamin (1995), in his book *Mushrooms: Poisons and Panaceas,* has provided a lucid interpretation of medicinal fungi from a medical standpoint. He summarized and criticized advances in research and applications and designated areas needing more investigation to assure meaningful outcomes.

MEDICINAL FUNGI OF SPECIAL INTEREST

Both traditional medicine (Ying et al., 1987) and scientific research (Jong and Birmingham, 1992, 1993) have shown that many, but not all, medicinal fungi that show much promise in the treatment of certain cancers and other pathologies are wood-rotting species in the orders Agaricales and Aphyllophorales of the phylum Basidiomycota (Alexopoulos et al., 1996; Gilbertson, 1980). Interestingly, the medicinal chemistry of these Basidiomycetes, especially that showing anticancer effects, involves polysaccharides (Jong and Birmingham, 1993; Kataoka-Shirasugi, 1994; Kiho and Ukai, 1995; Miyasaka et al., 1992; Shiao et al., 1994). It is possible that evolutionary adaptation to cellulosic and other complex carbohydrate substrates in wood-rot fungi is partly responsible for their versatility in polysaccharide synthesis. Whether such versatility has been reinforced evolutionarily by ecosystem forces is largely speculative at this time; however, emerging research involving fungal compounds as possible agents of ecosystem integration encourages such hypotheses (Jaenike, 1985). It is, of course, known that systems of human physiology, as elucidated below, "recognize" and respond in positive or negative ways to certain fungal chemicals.

Selected fungi are discussed below with regard to their production of medicinal and chemotherapeutic compounds. These examples were selected because of the relatively large amount of published, scientific research on their medicinal chemistry and actions, the progressive involvement of human subjects in research and trials, the relatively strong validation and corroboration of medicinal properties, increasing uses in medicine of certain fungal products, logical implications of certain chemistries on certain bodily systems, and well-established roles in traditional medicine. Table 2 shows selected medicinal chemical groups and their apparent therapeutic or other medicinal effects connected with these and other fungi. As in other areas of medical research, some reports of effects are based on studies with animals and may or may not be corroborated by trials on human subjects.

MEDICINAL CHEMISTRY OF *LENTINUS EDODES* (= *LENTINULA EDODES;* SHIITAKE)

This tawny, gilled mushroom (Figure 4), which grows on wood and which has been used as food and in medicine by the Chinese and Japanese for centuries, produces a variety of bioactive compounds (Jong and Birmingham, 1993). An old Japanese adage promises the addition of time to your life for every shiitake mushroom you eat. Scientific research to date has tended to reinforce the health-promoting nature of shiitake, although levels of validation vary according to the specific cases of chemistry and action considered. Shiitake is the most extensively studied of the medicinal fungi, and much corroborating data have been produced. The Japanese were pioneers in linking traditional knowledge with scientific research in this and later, other medicinal fungi. Jong and Birmingham have published an extensive review on the medicinal research on shiitake (Jong and Birmingham, 1993).

One of the chief attributes of shiitake is a β-1,3-glucan called lentinan, which has shown clear action as a general immune potentiator and antitumor agent (Jong and Birmingham, 1993). Lentinan is produced commercially and is a popular product in Japan. Other benefits (see Table 2) include hypolipidemic effects of an amino acid derivative eritadenine (= lentysine or lentinacin) (Jong and Birmingham, 1993; Sugiyama, 1993; Suzuki and Oshima, 1976), and antiviral peptidomannan (Jong and Birmingham, 1993), and nucleosides that are antithrombotic (Jong and Birmingham, 1993).

Dr. Eric Block called attention in this conference to the existence of organosulfur compounds, dipeptides containing cystein-*S*-oxides, in both *Marasmius* (garlic mushroom) and *L. edodes* (shiitake mushroom), that are similar to *Allium* compounds (Block, 1994; Block et al., 1994; Gmelin et al., 1976, 1980). As in *Allium,* some of these mushroom flavorants have shown

TABLE 2. Medicinal and Therapeutic Actions Associated with Selected Fungal Biochemicals Reported from Medical Research.

Biochemicals	Actions
Adenosine	Platelet aggregation inhibitor[a]
Alkaloids (*Claviceps*)	Vasoconstriction, etc.[b]
Amino acid derivative (Eritadenine, Shiitake)	Hypolipidemic[c]
β-Glucan (Lentinan in Shiitake)	General immune-potentiator; antitumor[d]
Dietary fiber (Shiitake)	Lower serum cholesterol[d]
Mushroom compounds	Antimutagen[e]
Nucleoside (*Cordycepin*)	Antibiotic[f]
	Antitumor[g]
(Shiitake)	Antithrombotic[d]
Organic acids (Ganodermic Acids)	Antihepatoxic[h]
	Hypocholesterolemic[h]
	Inhibit histamine release[h]
Organosulfur compounds (Shiitake)	Possibly useful antibiotics[i]
Other polysaccharides	Immunomodulation[j]
Aureobasidium (*Cordyceps*,	Antitumor[k]
Ganoderma, polypores)	Antitoxins[l]
	Hypoglycemic[m]
Oxygenated triterpernoids	Reduce cholesterol synthesis[h]
(Ganoderma)	Hypoglycemic activity[h]
	Cytotoxic to hepatoma cells[h]
Peptidogalactomannans (*Cordyceps*)	Antitumor[n]
Peptidomannan (Shiitake)	Promote interferon production[d]
Polypeptide (Ling Zhi-8, *Ganoderma*)	Antitumor[h]
	Immunomodulator[o]
	Immunosuppressant in grafts[p]
Taxol (*Taxomyces*)	Anti-ovarian cancer[q]

[a]Kawagishi, 1993.
[b]Agrios, 1997; Alexopoulos et al., 1996; Benjamin, 1995.
[c]Jong and Birmingham, 1993; Suzuki and Oshima, 1976.
[d]Jong and Birmingham, 1993.
[e]Grüter et al., 1990.
[f]Kuo et al., 1994.
[g]Kiho and Ukai, 1995; Kuo et al., 1994.
[h]Shiao, Lee, et al., 1994.
[i]Block, 1994; Block et al., 1994.
[j]Jong and Birmingham, 1993; Miyasaka et al., 1992.
[k]Kataoka-Shirasugi, 1994; Kiho and Ukai, 1995; Shiao, Lee, et al., 1994.
[l]Weete, 1974.
[m]Kiho and Ukai, 1995; Shiao, Lee, et al., 1994.
[n]Kiho and Ukai, 1995.
[o]Miyasaka et al., 1992.
[p]Vanderhem et al., 1994.
[q]Stierle, 1993.

Figure 4 *Lentinus edodes,* the Shiitake, a gilled mushroom that grows from rotting logs and is perhaps the most investigated medicinal fungus. [Illustrations from Ying et al. (1987), *Icons of Medicinal Fungi from Ching;* used with permission of Science Press, Beijing.]

antibiotic activity. Such compounds, as suggested by Dr. Block, could be additional, significant contributors to the health-promoting chemical repertoire of shiitake.

MEDICINAL CHEMISTRY OF *GANODERMA LUCIDUM* (LING ZHI OR REISHSI)

This is another wood-rotting Basidiomycete (Figure 5) that often produces mushroomlike fruiting bodies with glossy, reddish, slender stalks (stipes) and variously shaped, often eccentrically stalked, broad caps bearing pores instead of gills on their undersides (Alexopoulos et al., 1996; Ying et al., 1978). The stipe and cap contain the medicinals. This fungus also has been used as food and medicine by the Chinese, Japanese, and other Asians for centuries, and the folklore on its values is extensive (Ying et al., 1987). Su (1991) reviewed the taxonomy and physiologically active compounds of the genus *Ganoderma*. Jong and Birmingham (1992) reviewed medicinal and therapeutic values.

One extensively investigated component of *G. lucidum* is Ling-Zhi-8, a polypeptide capable of immunomodulation and other health-promoting effects (Miyasaka et al., 1992). It works by modulating adhesion (interaction) of molecules on immunocompetent cells. Stimulation of spleen cell meiosis has been noted. Immunosuppressive action has been noted in allograft survival (Vanderhem et al., 1994). Various oxygenated triterpenoids such as ganodermic acids and lucidemic acids have exhibited hypoglycemic activity, inhibited cholesterol synthesis, inhibited histamine release, and have been cytotoxic to hepatoma cells *in vitro* (Shiao et al., 1994). Adenosine from *Ganoderma* was found to inhibit platelet aggregation (Kawagishi, 1993). Biotechnologists have characterized Ling Zhi-8 (Jong and Birmingham, 1992; Shiao et al., 1994) and have cloned the gene encoding the peptide (Murasugi et al., 1991). Triterpenoids are being investigated for possible medicinal or other physioactive qualities. Strain specificity for triterpenoids has been studied in *G. lucidum* (Hirotana, 1993).

MEDICINAL CHEMISTRY OF *CORDYCEPS SPECIES* (CHAN HUA, SEMITAKE, TOCHUKASO)

This *Ascomycete* fungus (Figure 6) has also been called the "vegetable caterpillar." Actually, the fungus infects buried caterpillars or other insects, kills its prey, and in the following spring or summer sends up one or more 1- to 2-inch, slender, tapered, erect spore-bearing structures (stromata) which, in areas of crowded production, the Chinese have likened to a "grass" (Alexopoulos et al.,

Figure 5 *Ganoderma lucidum,* the Ling Zhi or Reishsi fungus; basidiocarps growing from buried wood. [Illustrations from Ying et al. (1987), *Icons of Medicinal Fungi from Ching;* used with permission of Science Press, Beijing.]

Figure 6 *Cordyceps sinensis;* ascorporic stromata growing from dead caterpillar hosts. [Illustrations from Ying et al. (1987), *Icons of Medicinal Fungi from Ching;* used with permission of Science Press, Beijing.]

159

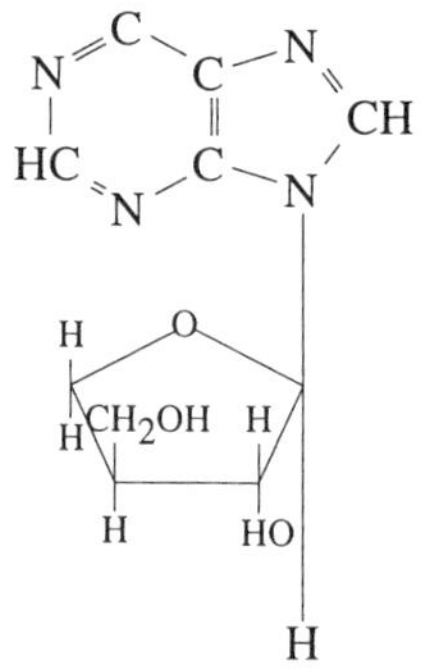

CORDYCEPIN

Figure 7 Cordycepin is a nucleoside antibiotic produced by the Ascomycete fungus *Codyceps,* a parasite of insects. Cordycepin has tumor-inhibiting properties and other potential effects.

1996; Ying et al., 1987). Medicinal compounds have been detected in both the infected caterpillar carcass, which is filled with fungus tissue, and the sporing fungal stroma of *C. sinensis* and *C. cicadae*. Compounds include the antitumor, antibiotic compound cordycepin, a nucleoside shown in Figure 7 (Kiho and Ukai, 1995; Kuo et al., 1994), and antitumor polysaccharides in *C. cicadae* (Kiho and Ukai, 1995). The antibiotic cordycepin may explain the use of *Cordyceps* in treating certain illnesses in Chinese medicine (Ying et al., 1987). Especially interesting is the scientifically reported amelioration by *Cordyceps* preparations of cyclosporine nephrotoxicity in kidney transplants (Xu et al., 1995). Many other claims of beneficial actions have been made for *Cordyceps,* but more research is needed to clarify and validate properties.

GENERAL NUTRITIONAL CHEMISTRY FROM FUNGI

Fungi in all three phyla are established or potential sources for many nutritional biochemicals, particularly those relating to amino acids, other organic acids, fatty acids, sugars, and vitamins (Beuchat, 1987; Garraway and Evans, 1984; Robinson, 1987). With developing biotechnologies, the list of new products as well as feasibilities for using fungal biochemicals are increasing. For example, strains of the yeast *Saccharomyces cerevisiae* have been engineered to produce lycopene (Yamano, 1994). Known and potential fungal contributions to the nutritional contents and character of food and food supplements are shown in Table 3.

There is strong interest in the production of fungal protein, and much research, development, and implementation work has been done (Sinskey and Batt, 1987). Exploitation of fungal protein production has been suppressed

TABLE 3. Principal and Potential Fungal Contributions to Foods and Food Supplements.

Contribution	Chemistry
Basic nutrition[a]	Fatty acids[b]
	Proteins, amino acids,[c] organic acids
	Sugars
	Vitamins, precursors[d]
Flavors/fragrances[e]	Alcohols, esters
	Isoprenoids
	Organosulfur compounds[f]
	Phenolics
Nondigestible bulk, modifiers, conditioners	Polysaccharides[g]
Potential bioactive, beneficial properties	Carotenoids[d]
	Lectins*,[h]
	Nucleosides*,[i]
	Phenolics*
	Polysaccharides[j]
	Triterpenoids*

*Indicates a relatively unexplored topic.
[a]Benchat, 1987; Robinson, 1987.
[b]Weete, 1974.
[c]Bigelis and Lasure, 1987.
[d]Yamano, 1994.
[e]Alexopoulos et al., 1996; Gallois et al., 1990; Gmelin et al., 1976, 1980; Pyysalo, 1975; Rapior, 1997.
[f]Block, 1994.
[g]Kataoka-Shirasugi, 1994; Nagata, 1993; Pollack, 1992; Schuster, 1993; Shiao, Wang, et al., 1994.
[h]Conrad, 1994; Kawagishi, 1997; Peumans, 1995.
[i]Shiao, Wang, et al., 1994.
[j]Jong and Birmingham, 1993; Kataoka-Shirasugi, 1994; Shiao, Lee, et al., 1994.
[k]Shiao, Lee, et al., 1994; Tai, 1993; Weete, 1974.

until now because of various feasibility problems, including substrate availability, chemical contamination, and markets. Some problems such as high ribonucleic acid content in the product can be overcome (Sinskey and Batt, 1987). The prospects for growth in this area would seem to be good because protein production constitutes a valuable process, among others, that could be integrated with fungal reduction of plant biomass, including such materials as municipal green waste, sugarcane bagasse (Pessoa, 1996), and sugar beet stillage (Lee, 1996). Emerging biotechnologies, including specially adapted yeasts and cross-fungal transplantation of specialized protein synthesis genomes (MacKenzie, 1993), promise to improve the feasibility and quality of production. Moreover, considering the versatility of yeasts for producing biochemicals (Beuchat, 1987), it is likely that other products will accrue from the same yeast biomass used for protein production. Yeast and yeastlike fungi such as *Candida* species, *Endomycopsis lipolytica,* and *Aureobasidium pullulans* as well as certain filamentous fungi such as *Trichoderma* have shown promise for protein production (Bigelis and Lasure, 1987; Cooke and Whipps, 1993; Lee, 1996; Pessoa, 1996).

FUNGAL CHEMISTRY SUPPORTING FOOD CHARACTERISTICS AND PREPARATIVE TECHNOLOGIES

In addition to contributing to nutritional content of foods, various fungi produce other substances that can contribute to the savory characteristics of food, utility of food additives, and to the formulary and preparative technologies of products. Chief among these contributions are chemicals affecting flavors and aromas, nondigestible bulk and textural qualities, and potentially healthful bioactive properties. Contributing chemical groups are shown in Table 3.

Much more can be done in the application of fungi to flavors and aromas. Fungi are recognized as some of the most fragrant and flavorful edible organisms (Alexopoulos et al., 1996; Pyysalo, 1975). Indeed, flavors and aromas in fungi may be species specific (Alexopoulos et al., 1996). Culture conditions may influence the formation of flavor compounds (Gallois et al., 1990). Granted, some fungi produce repulsive odors, but many others produce aromas and flavors reminiscent of almonds, fruits, bread, and other foods. Some fungi add unique, subtle, savory qualities to foods, possibly attributable to organic acids, carbohydrates, lipids, or other substances. The organosulfur compounds in shiitake and the garlic mushrooms are interestingly similar to flavorants in onions and garlic (Block, 1994; Block et al., 1994; Gmelin et al., 1976, 1980; Pyysalo, 1975; Rapior, 1997). More research is needed on fungal flavor and aroma chemistry, particularly in the areas of esters, phenolics, and isoprenoids.

Direct uses of mushrooms, morels, truffles, or other fungal "fruiting bodies" or tissues in soups, salads, and sauces are well established; however, in the future, their incorporation into many other products such as breads, condiments, and food supplements should be explored vigorously. This is especially true in the case of fruiting bodies of many higher Basidiomycetes in the orders Agaricales and Aphyllophorales (Alexopoulos et al., 1996), including shiitake, *Agaricus,* oyster (*Pleurotus* species), and the Chinese straw mushroom (*Volvariella volvacea*) (Chang, 1972) as well as some nonmushroom, wood-decaying fungi such as *Ganoderma lucidum,* the cauliflower fungus (*Sparassis*), and the stinkhorns *Phallus impudicus* and *Dictyophora indusiata* (Alexopoulos et al., 1996; Ying et al., 1987). These fungi offer distinctive, often subtle flavors or aromas and unique textural qualities, have proven edibilities, and most can be cultivated, stored, processed, and used in fresh, dried, or rehydrated states. Some could be pulverized for use as condiments or teas. There are even possibilities for application of sliced or flaked fungal preparations to the exterior of breads for visual, flavorful, and textural benefits or for the use of decoctions from fungi in beverages emphasizing natural ingredients. Many wild mushroom genera, such as *Boletus* and *Russula* (Alexopoulos et al., 1996), are known to have subtle to strong flavors and spicy or peppery characteristics, but some of these may not be suitable now for large-scale, commercial applications because they have not yet been cultivated or may have questionable edibilities for some people. Still they may be good candidates for research.

Much attention is being given to pullulan, a nondigestible exopolysaccharide (α-glucan) product of *Aureobasidium pullulans* that could be used in food, food processing, or in certain pharmaceutical products (Kataoka-Shirasugi, 1994; Nagata, 1993; Pollock, 1992; Robinson, 1987; Schuster, 1993). Applications include coating films (Shih, 1996), increased dispersibility in foods, textural modification, and alteration of food product components to improve utility. Rice protein can be combined with pullulans to produce edible films (Shih, 1996). Strains of *Aureobasidium* have been selected to reduce undesirable pigment in pullulan (Pollock, 1992).

FUTURE USES OF FUNGI TO EXPAND THE DIVERSITY AND UTILITY OF PHYTOCHEMICAL RESOURCES

MODIFICATION OF PHYTOCHEMICALS BY FUNGI

Many possibilities exist for modifying phytochemicals via fungal processing to obtain known (Beuchat, 1987) or novel compounds. This could be achieved through culturing of fungi with plant materials, use of purified fungal enzymes (Bigelis and Lasure, 1987) in reactive systems involving selected phytochemicals, or biochemical combinations of fungal and plant intermediates. Suitability of fungi for these purposes could be strengthened or tailored by genetic engineering and/or by conventional environmental and nutritional controls (Beuchat, 1987; Griffin, 1994; Onions et al., 1986).

Chemical candidates for modification could be phytochemicals that have already shown important attributes with regard to food qualities, food preservation, health-promoting effects, or other benefits. Likely fungi could be nonobligate (culturable) biotrophs such as parasites and postharvest rotters that have known interactive chemical relationships with substrate-source plants (Agrios, 1997) or saprotrophs (decomposers) with versatile enzymatic capabilities for altering plant residues (Cooke and Rayner, 1984; Cooke and Whipps, 1993). For example, certain fungi have many pathways for transforming phenolic compounds in plant tissues (Garraway and Evans, 1984;

Griffin, 1994). It would be interesting to take savory health-promoting phenolics from tea (*Camellia sinensis*), grapes (*Vitis*), or other sources and subject
them to fungi to create altered products that might exhibit stronger or different physioactive or savory qualities. Fungi such as the stone fruit rotter
Monilinia (Agrios, 1997), which is known for interactions with plant substrates leading to broad changes in phenolic metabolism, might be suitable
for such chemical-modifying services.

Other plant families and member crops or their products that offer biochemicals with known or potential nutritional or health-promoting qualities,
which possibly could be modified by fungi to form useful substances, are
listed in Table 4.

If filamentous fungi pose industrial handling problems, some of their genetic capabilities might in the future be biotechnically transferable to yeasts
or bacteria (MacKenzie, 1993).

Regardless of the procedural or production needs or concerns, whether they
be physical, chemical, or environmental, there are hundreds of adaptable fungi
to choose from and abundant literature to guide the researcher. Microbial
germplasm (culture) repositories around the world, such as the American Type
Culture Collection in Rockville, Maryland, are able to supply patented fungi as
well as others for use in research or production. Such repositories not only provide accurately identified microbes but may help to ensure their strain-specific
character and freedom from infectious agents, thereby assuring predictable per-

TABLE 4. Suggested Plant Families and Member Crops or Their Products
That Offer Biochemicals of Known or Potential Nutritional or Health-Promoting
Qualities That Could Be Experimentally Modified by Fungi to Produce
Possibly Useful Substances.

Family	Crop or Product	Biochemicals
Apiaceae	Celery relatives	Isoprenoids and derivatives; phenolics
Brassicaceae	Cole crops, seed oils	Carotenoids, thiols, isothiocyanates
Fabaceae	Bean curds	Proteins, polysaccharides
Myrtaceae	Eucalyptus, spices	Essential oils
Poaceae	Cereals	Polysaccharides
Rosaceae	Stone and pome fruits	Phenolics
Rutaceae	Citrus	Essential oils, organic acids
Theaceae	*Camellia* tea	Phenolics
Vitaceae	Grapes	Phenolics
Zingiberaceae	Ginger and relatives	Phenolics

formance. These "germplasm banks" may provide annotated catalogs or consultative services for selection of suitable fungi.

BIOMASS REDUCTION AND POTENTIAL CHEMICAL DERIVATIVES

Great and exciting challenges face us in the quest to reduce and/or recycle plant biomass. There are mounting pressures in different states and municipalities to reduce "green waste," and many industries are forming or retooling in response and are developing new ways to process this vast resource and make recycling operations profitable. Green waste could be broadly construed to include biomass not only from landscapes and nursery operations but also from marine kelp industries. Of course, there are many types of crop residues from agricultural production, postharvest processing activities, and from the manufacture of food supplements. Already many new strategies for dealing with biomass of various kinds have emerged; however, approaches so far have concentrated more on biomass volume reduction and composting applications than on chemical recycling for value-added derivatives or products. This presents a great opportunity for innovation by companies that are capable of tackling the more sophisticated, chemical side of the situation.

Since many fungi are easily cultivated, nonfastidious saprotrophs capable of growing vigorously on raw plant waste (Cooke, 1977; Ellis and Ellis, 1985), they are prime candidates of biomass reduction, including composting and potential biochemical recycling. Because of their enzymatic versatility (Garraway and Evans, 1984), fungi are vastly superior to bacteria in scope and speed of chemical degradation and volume reduction of raw plant waste—especially lignocellulosic material (Bonnen, 1994; Cooke and Whipps, 1993; Garraway and Evans, 1984; Thurston, 1994). Bacteria, of course, have their important contributions to biomass reduction and can be used in staged or cocultured microbial processing schemes.

Surely, there are special opportunities for the producer of plant foods and/or food supplements to find new ways to employ fungi in reducing waste while creating value-added products or services through integrated design of factory production objectives and operations. Figure 8 presents a general schematic for the manufacture of formulated products via the interaction of plant substrates or biomass with fungal processing.

In nature, some fungi can follow others in successive colonization and breakdown of the same substrate mass, such as decaying wood (Cooke and Rayner, 1984). Royce (1992) reported a practical application of this principal involving the recycling of spent shiitake culture substrate for the production of the oyster mushroom *Pleurotus sajor-caju*.

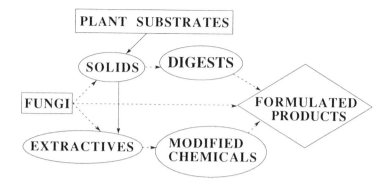

Figure 8 Suggested industrial system for manufacturing formulated products via interaction of phytochemicals or biomass with fungal processing

SACCHARIZATION AND FUNGAL PROCESSING

Liquid, slurry, and solid wastes such as sugarcane bagasse, recycled paper, coffee hulls, potato trimmings, cereal milling residues, and fruit culls can be subjected to acid hydrolysis (if necessary) and/or fungal digestion to release sugar from carbohydrate-rich debris, followed by fungal fermentation to create nutrient or "fiber"-containing mycobiomass and/or free chemical products (Beuchat, 1987; Cooke and Rayner, 1984; Onions et al., 1986). Such processes could be integrated with food or food supplement processes for both waste reduction and value-added chemical conversion purposes as suggested in Figure 8. The potential for coupling biomass reduction with fungal protein production was discussed previously, in the section General Nutritional Chemistry from Fungi.

Cellulases are produced by many Ascomycetes and Basidiomycetes, with *Trichoderma, Chaetomium,* and some wood-rotting basidiomycete genera having been most useful in industry, whereas pectinases are produced by various *Ascomycetes, Basidiomycetes, Oomycetes,* and *Zygomycetes* (Agrios, 1997; Bigelis and Lasure, 1987; Cooke and Whipps, 1993; Griffin, 1994; Onions et al., 1986). Many fungi are able to form a variety of these enzymes, thereby degrading various polysaccharides to different-sized fragments or to sugars. Products of degradation can be metabolized further to other biochemicals and fungal biomass.

Delignification of plant fiber by fungal digestion using such genera as *Phanerochaete* and *Ceriporiopsis* in the form of cellulase-less mutants improved digestibility by ruminants (Akin, 1993). This kind of technology may have potential applications to human food products and might be useful in freeing phenolic substances for further conversion and use.

The taxonomic groups of fungi useful in industry and insights to their handling have been reviewed by Onions et al. (1986) and Beuchat (1987). Alexopoulos (1996) is an excellent reference for general insights to, and illustrations of, all major fungal groups in nature and human affairs.

CONCLUSION

There is much to discover and profit by in our quest to jointly exploit two great resources: phytochemistry and the fungi. Research and development can be expected to greatly increase our stockpile of fungal compounds supporting the nutrition and health-related industries. The facilitation and new methodologies provided by biotechnology and improvements in existing technologies will increase the utility and productivity of fungi in food and pharmaceutical industries. Finally, imaginative exploitation of the interactive potentials between plant and fungal biochemistries will support biosynthesis and conversions leading to novel compounds, with significant implications for nutrition and health. These achievements will require scientists and technologists that have biotechnical skills backed up by broad, practical understandings of plants and fungi as well as the agricultural and industrial systems to be served by these achievements. These people will be flexible, holistic thinkers who can communicate effectively between disciplines and around the world and work well in teams to explore, solve problems, and create ideas that assure practical outcomes in one of the most essential and exciting fields in our world.

REFERENCES

Agrios, G. N. 1997. *Plant Pathology,* 4th ed. San Diego: Academic Press.

Akin, E. E. 1993. Microbial Delignification with White Rot Fungi Improves Forage Digestibility, *Appl. Environ. Microbiol.,* 59:4282–4282.

Alexopoulos, C. J., Mims, C. W., and Blackwell, M. 1996. *Introductory Mycology,* 4th ed. New York: John Wiley & Sons.

Benjamin, D. R. 1995. *Mushrooms: Poisons and Panaceas.* New York: W. H. Freeman & Co. 422 pp.

Beuchat, L. R., ed. 1987. *Food and Beverage Mycology,* 2nd ed. New York: Van Nostrand Reinhold.

Bigelis, R., and Lasure, L. L. 1987. Fungal Enzymes and Primary Metabolites Used in Food Processing, pp. 473–516, In Beuchat, L., ed., *Food and Beverage Mycology,* 2nd ed. New York: Van Nostrand Reinhold.

Block, E. 1994. Chemistry Is a Salad Bowl: Comparative Organosulfur Chemistry of Garlic, Onion and Shiitake Mushrooms, *Pure Appl. Chem.,* 66 (10–11): 2205–2206.

Block, E., DeOrazio, R., and Thiruvazhi, M. 1994. Simple Total Syntheses of Biologically Active Penthathiadecane Natural Products, 2,4,5,7,9-Pentathiadecane,

2,2,9,9-Tetraoxide (Disoxysulfone), from *Dysoxylum richii,* and 2,3,5,7,9-Pentathiadecane 9,9-Dioxide, the Misidentified Lenthionine Precursor SE-3 from Shiitake Mushroom (*Lentinus edodes*), *J. Org. Chem.,* 59(9):2273–2275.

Bonnen, A. M. 1994. Lignin-Degrading Enzymes of the Commercial Button Mushroom, *Agaricus bisporus, Appl. Environ. Microbiol.* 60:960–965.

Chang, S.-T. 1972. *The Chinese Mushroom.* Hong Kong: Chinese University of Hong Kong.

Christensen, C. M. 1975. *Molds, Mushrooms, and Mycotoxins.* Minneapolis: University of Minnesota Press.

Conrad, F. 1994. The Lectin from *Pleurotus ostreatus:* Purification, Characterization and Interaction with a Phosphatase, *Phytochemistry,* 36(2):277–283.

Cooke, R. 1977. *The Biology of Symbiotic Fungi.* New York: John Wiley & Sons.

Cooke, R. C., and Rayner, A. D. M. 1984. *Ecology of Saprotrophic Fungi.* New York: Longman Group, pp. 321–333.

Cooke, R. C., and Whipps, J. M. 1993. *Ecophysiology of Fungi.* Cambridge, MA: Blackwell Scientific Publications, pp. 23–43.

Davin, L. B., Wang, H.-B., Crowell, A. L., Bedgar, D. L., Martin, D. M., Sarkanen, S., and Lewis, N. G. 1997. Stereoselective Bimolecular Phenoxy Radical Coupling by an Auxiliary (Dirigent) Protein without an Active Center, *Science,* 275: 362–366.

Ellis, M. B., and Ellis, J. P. 1985. *Microfungi on Land Plants: an Identification Handbook.* New York: MacMillan.

Gallois, A., Gros, B., Langlois, D., Spinnler, H.-E., and Brunerie, P. 1990. Influence of Culture Conditions on Production of Flavor Compounds by 29 Ligninolytic Basidiomycetes, *Mycol. Res.,* 94:494–504.

Garraway, M. O., and Evans, R. C. 1984. *Fungal Nutrition and Physiology.* New York: John Wiley & Sons.

Geissman, T. A., and Crout, D. H. G. 1969. *Organic Chemistry of Secondary Plant Metabolism.* San Francisco: Freeman, Cooper & Co.

Gilbertson, R. L. 1980. Wood-Rotting Fungi of North America, *Mycologia,* 72:1–49.

Gmelin, R., N'Galamulume-Treves, M., and Hölfe, G. 1980. Epilentinsaure, ein Neuer Aroma und Geruchs-Precursor in *Tricholoma* Arten, *Phytochemistry,* 19: 553–557.

Gmelin, R., Luxa, H. H., Roth, K., and Hölfe, G. 1976. Dipeptide Precursor of Garlic Odour in *Marasmius* Species, *Phytochemistry,* 15:1717–1721.

Griffin, D. H. 1994. *Fungal Physiology,* 2nd ed. New York: Wiley-Liss.

Grüter, A. Friederich, U., and Wurgler, F. E. 1990. Antimutagenic Effects of Mushrooms, *Mutat. Res.,* 231(2):243–249.

Hirotana, M. 1993. Comparative Study of the Strain-Specific Triterpenoid Components of *Ganoderma lucidum, Phytochemistry,* 33(2):379–382.

Jaenike, J. 1985. Parasite Pressure and the Evolution of Amanitin Tolerance in *Drosophila,* Evolution, 39:1295–1301.

Jong, S. C., and Birmingham, J. M. 1992. Medicinal Benefits of the Mushroom *Ganoderma, Adv. Appl. Microbiol.,* 37(1):101–134.

Jong, S.C., and Birmingham, J. M. 1993. Medicinal and Therapeutic Value of the Shiitake Mushroom, *Adv. Appl. Microbiol.,* 39:153–184.

Kataoka-Shirasugi, N. 1994. Antitumor Activities and Immunological Properties of the Cell Wall Polysaccharides from *Aureobasidium pullulans, Biosci. Biotechnol. Biodiversity,* 58:2145–2151.

Kawagishi, H. 1993. 5'Deoxy-5'-Methylsulfinyl Adenosine, a Platelet Aggregation Inhibitor from *Ganoderma lucidum, Phytochemistry* 32(2):239–241.

Kawagishi, J. 1997. A Lectin from Mycelia of the Fungus *Ganoderma lucidum, Phytochemistry,* 44:7–10.

Kiho, S., and Ukai, S. 1995. Tochukaso (Semitake and Others), *Cordyceps* species, *Food Rev. Int.,* 11(1):231–234.

Kuo, Y.-C., Lin, C.-Y., Tsai, J.-J., Wu, C.-L., Chen, C.-F., and Shiao, M.-S. 1994. Growth Inhibitors Against Tumor Cells in *Cordyceps sinensis* Other Than Cordycepin and Polysaccharides, *Cancer Invest.,* 12(6):611–615.

Lee, K.-Y. 1996. Continuous Process for Yeast Biomass Production from Sugar Beet Stillage by a Novel Strain of *Candida rugosa* and Protein Profile of the Yeast, *J. Chem. Technol. Biotechnol.,* 66:349–354.

Lincoff, G., and Michell, D. H. 1977. *Toxic and Hallucinogenic Mushroom Poisoning.* New York: Van Nostrand Reinhold.

MacKenzie, D. A. 1993. Regulation of Secreted Protein Production by Filamentous Fungi: Recent Developments and Perspectives, *J. Gen. Microbiol.,* 139: 2295–2307.

Miyasaka, N., Inoue, H., Totsuka, T., Koike, R., Kino, K., and Tsunoo, H. 1992. An Immunomodulatory Protein, Ling Zhi-8, Facilitates Cellular Interaction Through Modulation of Adhesion Molecules, *Biochem. Biophys. Res. Commun.,* 186(1): 385–390.

Murasugi, A., Tanaka, S., Komiyama, N., Iwata, N., Kino, K., Tsunoo, H., and Sakama, S. 1991. Molecular Cloning of a cDNA and a Gene Encoding an Immunomodulatory Protein, Ling Zhi-8, from a fungus, *Ganoderma lucidum, J. Biol. Chem.,* 266(4):2486–2499.

Nagata, N. 1993. Fermentative Production of Poly (β-L-Malic Acid), a Polyelectrolytic Biopolyester, by *Aureobsidium* sp., *Biosci. Biotechnol. Biodiversity,* 57:638–642.

Onions, A. H. S., Allsopp, D., and Eggins, H. O. H. 1986. *Smith's Introduction to Industrial Mycology,* 7th ed. London: Edward Arnold.

Pessoa, A., Jr. 1996. Cultivation of *Candida tropicalis* in Sugar Cane Hemicellulosic Hydrolysate for Microbial Protein Production, *J. Biotechnol.,* 51:83–88.

Peumans, W. J. 1995. Lectins as Plant Defense Proteins, *Plant Physiol.,* 109:347–352.

Pollock, T. J. 1992. Isolation of New *Aureobasidium* Strains That Produce High Molecular-Weight Pullulan with Reduced Pigmentation, *Appl. Environ. Microbiol.,* 58:877–883.

Pyysalo, H. 1975. *Studies on the Volatile Compounds in Mushrooms.* Technical Research Center of Finland, Materials and Processing Technology Publication 13.

Quack, W., Anke, T., Oberwinkler, F., Gianneti, B. M., and Steglich. 1978. Antibiotics from *Basidiomycetes.* V. Merulidial, A New Antibiotic from the Basidiomycetes, *J. Antibiotics,* 31:737–741.

Rapior, S. 1997. Volatile Flavor Constituents of Fresh *Marasmium alliaceus* (Garlic Marasmius), *J. Agric. Food Chem.,* 45:820–825.

Robinson, D. S. 1987. *Food-Biochemistry and Nutritional Value.* New York: Longman/Wiley.

Robinson, T. 1980. *The Organic Constituents of Higher Plants: Their Chemistry and Interrelationships,* 4th ed. North Amherst, MA: Cordus Press.

Royce, D. J. 1992. Recycling of Spent Shiitake Substrate for Production of Oyster Mushroom, *Pleurotus sajor-caju, Appl. Microbiol. Biotechnol.,* 38:179–182.

Schuster, R. 1993. Production of the Fungal Exopolysaccharide Pullulan by Batchwise and Continuous Fermentation, *Appl. Microbiol. Biotechnol.,* 39:155–158.

Shiao, M. S., Lee, K. R., Lin, L.-J., and Wang, C.-T. 1994. Natural Products and Biological Activities of the Chinese Medicinal Fungus *Ganoderma lucidum, Am. Chem. Soc. Symp. Series,* 547:342–354.

Shiao, M. S., Wang, Z. N., Lin, L.-J., Lien, J.-Y., and Wang, J.-J. 1994. Profiles of Nucleosides and Nitrogen Bases in Chinese Medicinal Fungus *Cordyceps sinensis* and Related Species, *Botanical Bull. Acad. Sinica* 35(4):261–267.

Shih, F. 1996. Edible Films from Rice Protein Concentrate and Pullulan, *Cereal Chem.,* 73:406–409.

Sinskey, A. J., and Batt, C. A. 1987. Fungi as a Source of Protein, In Beuchat, L. R., ed., *Food and Beverage Mycology,* 2nd ed. New York: Van Nostrand Reinhold, pp. 435–471.

Stamets, P. 1993. *Growing Gourmet and Medicinal Mushrooms.* Berkeley: Ten Speed Press.

Stamets, P. 1996. *Psilocybin Mushrooms of the World: An Identification Guide.* Berkeley: Ten Speed Press, pp. 1–16.

Stierle, A. 1993. Taxol and Taxane Production by *Taxomyces andreanae,* an Endophytic Fungus of Pacific Yew, *Science,* 260:214–216.

Stoner, M. F. 1994. Consideration of Agroecosystems is Essential in Holistic Environmental Conservation and Management, pp. 195–197, In *Proceedings of the 22nd Annual Conference of the North American Association for Environmental Education,* Sept. 25–28, 1993. Big Sky, Montana.

Stoner, M. F., and Baker, G. E. 1981. Soil and Leaf Fungi, pp. 171–180. In D. Meuller-Dombois et al., ed., *Island Ecosystems: Biological Organization in Selected Hawaiian Communities.* US/IBP Synthesis Series 15. Stroudsburg, PA: Hutchinson Ross.

Su, C.-H. 1991. Taxonomy and Physiological Active Compounds of *Ganoderma*—A Review, *Bull. Taipei Med. Coll.,* 20:1–16.

Sugiyama, K. 1993. The Hypocholesterolemic Action of *Lentinus edodes* Is Evoked Through the Alteration of Phospholipid Composition of Liver Microsomes in Rats, *Biosci. Biotechnol. Biochem.,* 57:1983–1985.

Suzuki, S., and Oshima, S. 1976. Influence of Shii-ta-ke *(Lentinus edodes)* on Human Serum Cholesterol, *Mushroom Sci.,* 9(2):463–467.

Tai, T. 1993. Triterpenes from *Poria cocos, Phytochemistry* 32(5):1239–1244.

Thurston, C. F. 1994. The Structure and Function of Fungal Laccases, *Microbiology,* 140:19–26.

Vanderhem, L. G., Vandervliet, J. A., Bocken, C. F. M., Kino, K., Hoitsma, A. J., and Tax, W. J. M. 1994. Prolongation of Allograft Survival with Ling Zhi-8, a New Immunosuppressive Drug, *Transplant Proc.,* 26(2):746.

Weete, J. D. 1974. *Fungal Lipid Biochemistry.* New York: Plenum Press.

Xu, F., Huang, J. B., Jiang, L., Xu, J., and Mi, J. 1995. Amelioration of Cyclosporine

Nephrotoxicity by *Cordyceps sinensis* in Kidney-Transplanted Recipients, *Nephrol. Dialysis Transplant.*, 10(1):142–143.

Yamano, S. 1994. Metabolic Engineering for Production of β-Carotene and Lycopene in *Saccharomyces cerevisiae, Biosci. Biotechnol. Biochem.*, 58:1112–11114.

Ying, J., Mao, X., Ma, Q., Zong, Y., and Wen, H. 1987. *Icons of Medicinal Fungi from China.* Beijing: Sciences Press.

Zhang, X-D. 1997. Phenoloxidases in Portabello Mushrooms, *J. Food Sci.*, 62:97–100.

CHAPTER 10

Developing Claims for New Phytochemical Products

RICHARD E. LITOV

INTRODUCTION

THE relatively new term, *phytochemical,* which broadly defined is a plant-based substance, is appearing more frequently in the media and scientific literature. However, plant substances have been used from before recorded history by shamans and medicine men to treat the multitude of ailments that afflict humankind. Even today, 80% of the world's population use medicinal plants to meet their primary health care needs (Bannersman, 1982). In the United States, 25% of prescriptions are plant-derived drugs (Srivastava et al., 1996). Currently, there is high interest in adding bioactive phytochemicals to foods or in formulating them into dietary supplements. These new products are being developed to go beyond meeting the basic nutritional needs of growth and maintenance; they are expected to provide additional health benefits. The targeted benefits include preventing disease, enhancing physical performance, and improving the quality of life.

Health messages for new phytochemical products can be made depending on what ingredients and levels are used and how the product is positioned. An overview of the current regulatory categories under which such products can be positioned and the limitations and allowances for health messages within these categories is given.

173

CONSUMER ATTITUDES ABOUT FOOD LABELS

Increasingly, consumers recognize the connection between diet and disease, and many are willing to change their eating habits. The U.S. Food and Drug Administration (FDA) conducted a national telephone survey of consumers to monitor the impact of the recent changes in food label regulations (Levy and Derby, 1996). When asked about the accuracy of label claims, only 31% of consumers said "about all" or "most" health claims are accurate. Thus, there is widespread skepticism about health claims made on food labels.

REGULATORY CATEGORIES FOR PHYTOCHEMICALS

There are already numerous phytochemical products on the market and many more are in development. Currently, a new phytochemical product may be eligible for regulatory status as a (1) food, (2) dietary supplement, or (3) medical food. The basic differences among these regulatory categories are highlighted below. Deciding which regulatory route to follow is an important early step in developing new phytochemical products. This choice depends on many factors, including business goals and available resources. It should be realized that there are many areas within these regulations that are subject to interpretation. Much of the current uncertainty will resolve as the FDA begins to establish a clearer pattern of enforcement and as case law emerges.

FOOD

The Nutrition Labeling and Education Act (NLEA) of 1990 (U.S. Congress, 1990) for foods allows for 3 types of claims: (1) nutrient content, (2) structure-function, and (3) health. Nutrient content claims characterize the level of a nutrient, based on a predetermined reference amount of the food product. Claims are permitted only for those nutrients listed in the NLEA and implementing regulations. Although allowed, structure-function claims are rare or not currently used by food marketers. Nine health claims have been approved and are used with some food products. However, they have several limitations, including that the model claim language is prescribed with no mention of product brand allowed. Petitions for new health claims can be submitted, but getting approval is long, difficult, and costly.

DIETARY SUPPLEMENTS

The Dietary Supplement Health and Education Act of 1994 (U.S. Congress, 1994) repositions a product as different from food, exempts it from food additive provisions, and allows for flexible product claims. A product must be labeled as a dietary supplement and cannot be represented as a meal replacement or a conventional food. Unlike foods, dietary supplements are allowed to use "nutritional support statements." There is no formal approval process, manufacturers only have to notify the FDA of the statement's use within 30 days after first marketing the product. There are 4 types of "nutritional support statements": (1) classical nutrient-deficiency disease support (e.g., vitamin C and scurvy), (2) description of the intended role that the ingredient plays in affecting structure or function of the body, (3) characterization of the mechanism by which the ingredient acts to maintain such structure or function, and (4) description of the effect on general well-being expected from consumption of the ingredient. Marketers may not make any reference to disease conditions when using these statements in labeling and advertising. In addition, the following must be prominently displayed on the label: "This statement has not been evaluated by the Food and Drug Administration. This product is not intended to diagnose, treat, cure or prevent any disease."

Although dietary supplements are exempt from the food additive regulations, they are subject to a separate safety standard. New ingredients, those first marketed after October 15, 1994, must be "reasonably expected to be safe" when used under conditions recommended or suggested on the label. A bibliography of published articles on the new ingredient's safety must be provided to the Secretary of Health and Human Services 75 days before market introduction.

MEDICAL FOOD

Medical foods are the least formally regulated of the 3 categories. A definition of medical foods (U.S. Congress, 1990) is as follows: "A medical food is a food which is formulated to be consumed or administered enterally under the supervision of a physician and which is intended for the specific dietary management of a disease or condition for which distinctive nutritional requirements, based on recognized scientific principles, are established by medical evaluation." Additional regulations are expected as an advance notice of proposed rulemaking was made in November of 1996 (Food and Drug Administration, 1996). Traditionally, these products have been promoted through health care professionals and distributed to hospitals, nursing homes, and home health care with retail distribution in pharmacy settings.

REFERENCES

1. Bannerman, R. H. 1982. Traditional medicine in modern health care, *World Health Forum,* 3(1):8–13.
2. Srivastava, J., Lambert, J., and Vietmeyer, N. 1996. Medicinal Plants, World Bank Technical Paper #320, 1–21.

3. Levy, A. S., and Derby, B. M. January 23, 1996. The Impact of the NLEA on Consumers: Recent Findings from FDA's Food Label and Nutrition Tracking System, Internal Report of the Food and Drug Administration, Consumer Studies Branch, Center for Food Safety and Applied Nutrition.

4. U.S. Congress. 1990. Nutrition Labeling and Education Act of 1990, Public Law No. 101–535.

5. U.S. Congress. 1994. Dietary Supplement Health and Education Act of 1994, Public Law No. 103–417.

6. Food and Drug Administration. 1996. *Federal Register,* 61(231):60661–60671.

Index